BIOMATERIALS RESEARCH ADVANCES

BIOMATERIALS RESEARCH ADVANCES

JASON B. KENDALL
EDITOR

Nova Science Publishers, Inc.
New York

For permission to use material from this book please contact us:
Telephone 631-231-7269; Fax 631-231-8175
Web Site: http://www.novapublishers.com

LIBRARY OF CONGRESS CATALOGING-IN-PUBLICATION DATA
Biomaterials research advances / Jason B. Kendall (editor).
p.; cm
Includes bibliographical references.
ISBN-13: 978-1-60021-892-7 (hardcover)
ISBN-10: 1-60021-892-7 (hardcover)
1. Biomedical materials. I. Kendall, Jason B.
[DNLM: 1. Biocompatible Materials. 2. Nanostructures. 3. Tissue Engineering. QT 37 B5755 2008]
R857. M3B57344 2008
610.28--dc22

2007032047

Published by Nova Science Publishers, Inc. ✦ New York

CONTENTS

PREFACE

Biomaterials serve as synthetic or natural materials used to replace parts of living systems or to function contact with living tissue. Biomaterials are intended to interface with biological systems to evaluate, treat, augment or replace any tissue, organ or function of the body. A biomaterial is different from a biological material such as bone that is produced by a biological system. Artificial hips, vascular-stents, artificial pacemakers, and catheters are all made from different biomaterials and comprise different medical devices. This book presents new approaches to biomaterial development including multi-field bone remodeling, novel strategies for conferring antibacterial properties to bone cement, polyacrylonitrile-based biomaterials for enzyme immobilization and functionalized magnetic nanoparticles for tissue engineering from around the globe.

Chapter 1 - The objective of this review is to summarize basic mechanisms through which natural hemostatic systems interact with foreign materials. The strong activation of hemostatic systems is advantageous when a foreign material is meant to provide surface (topical) hemostasis for bleeding cessation. Alternatively, for many applications a benign interaction is desirable when foreign materials are in long-term contact with blood. Examples of this latter category are numerous, and include *in vivo* applications such as prosthetic heart valves or *in vitro* applications, *e.g.* blood collection tubes. This chapter describes basic principles learned from in-depth analysis of the interactions of a few select materials with hemostatic systems, which may be subsequently applied for the rational design of medical products. The goal is not to cover the extensive phenomenological literature that documents how different foreign materials interact with hemostatic systems. The response of hemostatic systems to artificial surfaces is hypothesized to occur in three stages: An initial selective adsorption event that is a function of a material's surface structure and chemical properties; a conformational distortion of the adsorbed proteins; and a functional reaction that is a consequence of the conformational alterations of the adsorbed proteins. This review examines biophysical mechanisms through which plasma proteins undergo chemical and physical adsorption to foreign materials, and how platelets and the proteins of the intrinsic coagulation pathway subsequently interact with the proteins that bind to the foreign materials. The interaction of hemostatic systems with poly-*N*-acetylglucosamine (pGlcNAc) nanofibers is examined in detail to illustrate how the aforementioned three-step interaction process functions.

Chapter 2 - The interactions of cells with biomaterials have been widely studied. However, little is known about the influence of the properties and chemistry of the substratum on the activation of one of the major signaling cascades, the cyclic AMP (cAMP) pathway, in adhering cells. The second messenger cAMP plays a major role in modulating cell morphology, enabling cells to survive, proliferate and differentiate. The authors previous study showed that murine Swiss 3T3 fibroblasts loosely attached to a hydrophilic cellulose membrane have a high amount of intracellular cAMP. By contrast, cells well spread on tissue culture polystyrene (PS) contained low concentrations of cAMP. But the surface properties of the cellulose membrane are heterogeneous in terms of rugosity, porosity and chemical composition.

Hence, to study the impact of surface chemistry on Swiss 3T3 fibroblast behaviors, the authors have prepared self assembled monolayers (SAMs) on glass from alkylsilanes to obtain model surfaces with a variety of terminating functional groups, such as carboxylic acid (COOH), amine (NH_2), poly(ethylene glycol) (PEG) and methyl (CH_3). The authors results revealed that the cAMP production was significantly lower in cells attached to COOH- and NH_2-terminated SAMs than in cells on PEG and CH_3 substrata. Spread cells attached to COOH- and NH_2-terminated SAMs could organize their cytoskeleton, phosphorylate the Tyr^{397} of focal adhesion kinase (FAK) and activate RhoA. By contrast, cells on PEG and CH_3 substrata remained rounded up with few punctuate focal adhesion complexes, while FAK phosphorylation on Tyr^{397} and RhoA activation were partly inhibited. Indeed, an increase in intracellular cAMP severely impaired the formation of focal adhesion complexes and decreased the phosphorylation of Tyr^{397} in FAK. The authors also found that the phosphorylation of ERK1/2 was also significantly greater in cells attached to COOH and NH_2 substrata than in cells adhering to PEG-terminated SAM. We conclude that monitoring cAMP may contribute to a better understanding of the complex phenomenon of cell-material interactions.

Chapter 3 - Endothelial cells (EC) play a vital role in tissue engineering (TE) - ranging from the design of small- to medium-sized tissue engineered blood vessel (TEBV) constructs to the creation of micro-vascular networks essential for the supply of oxygen and nutrients to the three-dimensional (3D) tissue assemblies. The first part of the study compared the behavior of bovine aortic arterial endothelial cells (BEC) cultured in a 3D gelatin scaffold having two different pore sizes, with that of the conventional 2D culture. DNA assay, PI staining, SEM and RT-PCR were done to evaluate the behavior of EC in 3D culture conditions. Specific emphasis was laid on the effect of pore size on EC behavior. The second part of the study evaluated BEC following treatment with low energy laser irradiation (LELI) from a diode laser. Recent work has focused on enhancing EC functions by the physical stimulation such as cyclic mechanical stress. The relatively few studies on the effect of low energy laser irradiation (LELI) on EC have been mostly been done on venous EC and have used He-Ne (helium neon) lasers. So BEC cultures were treated with LELI- having different energies and for different time periods. MTT tests, propidium iodide (PI) staining followed by FACS analysis and RT-PCR tests were done to determine cell viability, cell-cycle profiles and endothelial nitric oxide synthase (eNOS) gene expression. An increase in the proliferation rates and gene expression was observed at certain specific intensities. The comparative study of EC in 3D and 2D cultures in this study may provide some valuable background information that might useful in the future evaluation of EC for TE applications. An increase in the eNOS gene expression following LELI may be of potential benefit in the use of EC for various applications especially for EC in 3D biomaterials

Chapter 4 - The potential utility of polymer based resorbable implants in structural support applications, as biological container, and protective adhesion barrier in spinal surgery has been the focus of research in recent times. Accumulated preclinical experience coupled with improved polymer chemistry has allowed the clinical introduction of these devices into spinal surgery. The main focus of research has been on implants intended to aid spinal fusion (cages, screws, rods, scaffold carriers) and protective adhesion barriers for the neural elements.

Resorbable fusion implants are manufactured from polymers of which polylactic acid (PLA) is the most significant component. Preclinical studies have demonstrated adequate biocompatibility, sufficient stiffness and strength whilst load transfer to healing graft bone is gradual. Preclinical studies have also indicated that several parameters, including crystallinity, molecular weight, implant design and method of sterilization, which all affect the physical and biological properties of PLA based implants, may influence their clinical performance. The most clinical experience have been gained using the co-polymer 70:30 PLDLLA cage implant with which 87-97% spinal fusion rates have been reported, rates comparable to that seen with routinely applied non-resorbable implants. Also a novel application of a resorbable film as a protective adhesion barrier to neural elements has been reported with promising results.

The clinical success of various PLA based implants in spinal surgery seems to be influenced by the anatomical site and the corresponding local physiological environment. The mode of failure of PLA based implants is time dependent and is influenced by the nature of static and dynamic loading in vivo. Consequently future research should be directed towards models that will help understand and predict the biological and biomechanical behavior and performance of these polymers. Furthermore, there is need to clarify the influence of anatomical site of implantation on the behavior of the implant, with a goal towards the development of site specific implants.

Chapter 5 - The improvement of the biological activity and performance of bone substitute materials is one of the main concerns of orthopaedic and dental surgery specialists. Biomimetic nanocrystalline apatites exhibit enhanced and tunable reactivity as well as original surface properties related to their composition and mode of formation. Synthetic nanocrystalline apatites analogous to bone mineral can be easily prepared in aqueous media and one of their most interesting characteristics is the existence of a hydrated surface layer containing labile ionic species. Ion exchange and macromolecule adsorption processes can easily and rapidly take place due to strong interactions with the surrounding fluids. The ion mobility in the hydrated layer allows direct crystal-crystal or crystal-substrate bonding. The fine characterization of these very reactive nanocrystals is essential and can be accomplished with different tools including chemical analysis and spectroscopic techniques such as FTIR, Raman and solid state NMR. The reactivity of the hydrated layer of apatite nanocrystals offers material scientists and medical engineers extensive possibilities for the design of biomaterials with improved bioactivity using unconventional processing. Indeed apatitic biomaterials can be processed at low temperature which preserves their surface reactivity and biological properties. They can also be associated in various ways with active molecules and/or ions. Several examples of use and processing of nanocrystalline apatites involved in the preparation of tissue-engineered biomaterials, cements, ceramics, composites and coatings on metal prostheses are presented.

Chapter 6 - Biological and synthetic hydroxyapatites (HAp) display crystal structure similarities and differences that affect greatly the bioactivity of the synthetic materials. Crystal structure studies from x-ray (XPD) and neutron powder diffraction (NPD) of HAp based biomaterials is discussed in this chapter.

First, a comparison of structural parameters of natural and synthetic apatites from Rietveld refinements of high-resolution NPD patterns as a function of temperature is presented. The natural apatite samples are a carbonate fluorapatite (francolite) and a fluorapatite (harding pegmatite); the synthetic ones are low temperature HAp, and carbonated HAps. Modification of the structural parameters due to the carbonate substitution show a systematic behavior that is consistent with the mechanism of carbonate substitution on the mirror plane of the phosphate tetrahedron.

Then, the effect of silicon substitution on the crystal structure parameters of HAp is discussed from Rietveld refinement analysis of high-resolution NPD patterns as a function of temperature from samples of pure and 0.4 wt % silicon substituted HAp. Small structural changes in the lattice constants, interatomic distances, site occupancies and distortion of the phosphate tetrahedron were found.

In the third part, the structural and magnetic properties of ferrimagnetic bioglass ceramics (FBC) in the system $[0.45(CaO,P_2O_5)(0.52\text{-}x)SiO_2\ xFe_2O_3\ 0.03Na_2O]$, x=0.05, 0.10, 0.15, 0.20, as prepared and after heat treatment in the temperature range 600-1100 °C are assessed. Structure and microstructure of the materials as a function of temperature are studied using x-ray diffraction, scanning electron microscopy, and energy dispersive x-ray spectroscopy. The magnetic properties of FBC are correlated with their bulk and surface structure.

Finally, experimental results of the effect of simulated body fluids (SBF) on the crystal structure and microstructure of FBC are presented.

Chapter 7 - Tissue engineering represents a biology driven approach by which biological tissues are engineered through combining material technology and biotechnology. Its advantage over other tissue replacement techniques are several, *e.g.* use of autologous cells, nonimmunogenecity, no side-effects related to foreign graft materials, and potential to grow when implanted into children. Autologous cells of the tissue recipient are seeded on matrices that are fashioned from natural materials, or from synthetic polymers. The cell-matrix constructs are cultured *in vitro* to constitute a bioartificial tissue, the engineered implant for reconstructive surgery. In *in vitro* applications, bioartificial tissues serve as test systems for pharmaceutical drug screening and patient specific therapy. However, tissue engineering of complex tissues and organs is limited by their need of a vascular supply to guaranty graft survival and render bioartificial organ function. Therefore numerous strategies have been developed to overcome this hurdle including indirect revascularization, the concept of wrapping the generated graft with viable tissue, and stimulating ingrowth of microvessels by angiogenic factors, cells and stem cells. The development of a primary vascularized biological scaffold providing a vascular tree including a capillary network for the engineered implant may afford vascular anastomosis of any bioartificial construct to the recipient blood supply.

In: Biomaterials Research Advances
Editor: J. B. Kendall, pp. 1-19
ISBN: 978-1-60021-892-7

Chapter 1

MECHANISMS FOR THE INTERACTION OF HEMOSTATIC SYSTEMS WITH FOREIGN MATERIALS

Thomas H. Fischer, Carr J. Smith and John N. Vournakis

ABSTRACT

The objective of this review is to summarize basic mechanisms through which natural hemostatic systems interact with foreign materials. The strong activation of hemostatic systems is advantageous when a foreign material is meant to provide surface (topical) hemostasis for bleeding cessation. Alternatively, for many applications a benign interaction is desirable when foreign materials are in long-term contact with blood. Examples of this latter category are numerous, and include *in vivo* applications such as prosthetic heart valves or *in vitro* applications, *e.g.* blood collection tubes. This chapter describes basic principles learned from in-depth analysis of the interactions of a few select materials with hemostatic systems, which may be subsequently applied for the rational design of medical products. The goal is not to cover the extensive phenomenological literature that documents how different foreign materials interact with hemostatic systems. The response of hemostatic systems to artificial surfaces is hypothesized to occur in three stages: An initial selective adsorption event that is a function of a material's surface structure and chemical properties; a conformational distortion of the adsorbed proteins; and a functional reaction that is a consequence of the conformational alterations of the adsorbed proteins. This review examines biophysical mechanisms through which plasma proteins undergo chemical and physical adsorption to foreign materials, and how platelets and the proteins of the intrinsic coagulation pathway subsequently interact with the proteins that bind to the foreign materials. The interaction of hemostatic systems with poly-*N*-acetylglucosamine (pGlcNAc) nanofibers is examined in detail to illustrate how the aforementioned three-step interaction process functions.

INTRODUCTION

The development of medical devices, which by definition create biological/artificial surface interfaces, represents a significant current trend in biotechnology. This trend will accelerate as more nanotechnologies are advanced into the human clinical setting. However, in contrast with many areas of biotechnology where dramatic increases in understanding have taken place, such as in our ability to manipulate and use genes and genetic information, knowledge of how biological systems interact with foreign materials has advanced in a less dramatic fashion. A vast array of new materials with biomedical applications has been developed, but understanding of processes at the tissue/artificial surface interface is largely limited to phenomenological observation. The degree to which foreign materials activate hemostatic systems has been extensively documented (a recent PubMed search for articles that investigate hemostatic system activation by foreign surfaces yielded over 2,000 citations), but the supporting studies are usually limited to reporting that a given material does or does not activate platelets and/or humoral coagulation and/or adsorb plasma proteins (see Packham1 for a review). What is missing is a detailed mechanistic understanding of how foreign materials affect hemostatic systems. What are the relevant surface receptor systems that drive platelet activation on some foreign surfaces but not others? Why do certain surfaces and not others, adsorb plasma proteins and what is the effect of the adsorption on protein structure? These are the types of fundamental mechanistic questions addressed in this review. A result of our lack of understanding of the relationship between material science and hematology is that material design aspects in medical device development are largely based on an empirical trial and error approach. Our understanding of artificial surface-hemostatic system interactions has evolved to include three mechanistic steps: Adsorption of plasma proteins to the artificial surface; conformational distortion of the adsorbed plasma proteins; and activation of response of elements of the hemostatic system (e.g., platelets, intrinsic coagulation cascade).

OBSERVATIONS THAT DEFINE THE THREE-STEP PROTEIN ADSORPTION, CONFORMATIONAL DISTORTION, AND SYSTEM ACTIVATION INTERACTION MECHANISM

The importance of hemostatic system activation when blood contacts foreign materials is part of recorded human experience. The use of cloth bandages for achieving surface (topical) hemostasis is referred to in the ancient Greek literature [2] and Roman soldiers utilized chitin preparations for hemorrhage control. The discovery between 1840 and 1920 of platelets [3-5], the humoral coagulation system [6], and the concerted function of these aspects of hemostasis [7,8], laid the groundwork for seminal mechanistic findings concerning *how* hemostatic systems are activated when blood contacts artificial surfaces. The first observations concerned the activating response of elements of the hemostatic system to foreign materials, *i.e.* the last step of the three-step interaction mechanism indicated above. Platelets were found to adhere to glass and as a consequence, undergo an activation response[9]. Coagulation factor XII (FXII or Hageman factor) was subsequently discovered and found to be important in initiating humoral coagulation at the glass/blood interface [10,11] These initial findings were followed by studies which characterized how plasma proteins[12-14], including FXII [11] and fibrinogen [15-17], undergo chemical and physical adsorption processes on foreign surfaces [18]; the first step of the three-step interaction mechanism.

After the initial discoveries that factor XII and fibrinogen bind to artificial surfaces for hemostatic system activation, a great deal of work has focused on understanding how the adsorption event causes conformational distortions of the plasma proteins; the middle step in the three-step interaction mechanism. Extensive investigation has focused on how adsorbed fibrinogen [12-14,19-22] activates platelets and bound FXII [23,24] turns over the kallikrein/kinin contact and intrinsic coagulation systems. Artificial surface-bound fibrinogen has been found to behave like a "biosensor" for platelets. The adsorption of fibrinogen to glass is associated with an enthalpy that has been determined by calorimetry [25]. The material-bound fibrinogen conformation is inferred from biophysical methods of analysis such as atomic force microscopy [15], antibody epitope mapping [26], circular dichroism, Fourier transform infrared spectroscopy, and intrinsic fluorescence (see Heitz and Van Mau for a review of biophysical methods for adsorbed protein analysis) [27]. NMR methods that yield atomic-scale information, *e.g.* G-matrix Fourier transform [28,29] or double quantum [30] NMR with magic angle spinning, are only now being applied to small model systems [31,32]. Surface bound fibrinogen reportedly undergoes a poorly defined process termed "conversion" [33], that may involve disassociation from the surface, a conformational change and/or occlusion by prekallikrein or another protein [26].

The conformational distortion of fibrinogen on glass [34] and other hydrocarbon polymer-based materials [35-37] results in platelet adhesion and activation. A general pattern that has emerged from hemocompatibility experimentation (see Wang *et al.*) [38] is that less plasma protein, including fibrinogen, is adsorbed to interfaces that are chemically similar to the native plasma environment of fibrinogen, and there is a reduced tendency to activate hemostatic systems. For example, polyethyleneglycol (PEG) coated materials present a water-rich polar interface to plasma proteins for reduced adsorption and platelet activation [38]. Similarly, interfaces that resemble the neutral phospholipid bilayers that fibrinogen randomly contacts *in vivo* tend to be benign with respect to adsorption and activation [39]. Current understanding of integrin outside-in signaling processes suggests that integrins bind to domain(s) on the adsorbed fibrinogen molecule resembling fibrin, and then cluster on the platelet membrane to organize cytoskeletal-related signaling machinery for activation of outside-in signaling [40]. Many features of the fibrinogen adsorption, integrin activation processes are poorly understood. It is not known which fibrinogen isoform(s) adsorb and mediate the integrin activation response. Understanding how fibrinogen conformation and integrin activation affect the hemostatic properties of artificial surfaces is a key, but incompletely understood, element in the rational design of foreign materials.

The role of FXII in hemostatic system activation has been enigmatic because of the apparently "normal" hematological phenotype of patients lacking this coagulation factor [10]. FXI, but not FXII deficiency, results in prolonged bleeding [41]. This result indicates that alternative mechanisms are operant in intrinsic coagulation pathway activation. Several observations have led to the hypothesis that FXII can function as an anticoagulant (*e.g.*, see the review by Colman) [42]. This factor has an incompletely defined catalytic relationship with prekallikrein; FXII can proteolyze the kallikrein zymogen to the active kallikrein, which in turn can initiate hydrolysis of kinins to release the bradykinin peptide (see Colman for review) [42]. Related to the turnover of the kallikrein/kinin contact system is the stimulation of fibrinolysis. The combination of bradykinin-mediated vasodilation and fibrinolytic system activation defines a potential anti-hemostatic role for FXII. However, FXII (-/-) transgenic knockout mice have a reduced tendency for platelet thrombi to propagate and stabilize beyond points of initial platelet adhesion in an arterial endothelial injury [43]. These findings suggest that the effect of FXII on hemostatic balance might be highly tissue specific in a manner that is poorly understood.

While the activation of intrinsic coagulation by artificial surfaces has been recognized since the discovery of Hageman factor [11] and the original definition of the coagulation cascade, the underlying mechanism for FXII activation is only partially defined. Binding of FXII to foreign materials has been shown to involve *N*-terminus domains iso1-cys28 [44] and tyr135-arg153 [45]. The result of binding can be a conformational change [46-48] that facilitates hydrolysis at arg153 for activation by kallikrein [49-53], or by autoactivation of FXII [54] (see Cochrane and Griffin for a review) [55]. Hageman factor activation is balanced by several factors that appear to inhibit the hydrolytic activity of this protein [56,57], including C1q [58] and C1-inhibitor [59]. While many materials have been shown to active FXII to differing extents (kaolin, glass, *etc.*)[54,60-62],the molecular factors for specificity differences are poorly understood. While the degree to which various surfaces activate FXII can be a function of surface charge, more detailed molecular factors are probably involved. For example, we have shown that poly-*N*-acetylglycosamine (pGlcNAc) molecules that are self-associated in fibers in an anti-parallel orientation (β-conformation) activate intrinsic coagulation in a much more efficient manner that when the polymers are oriented in parallel to each other (α-conformation) [63]. This type of observation suggests that FXII activation is driven by specific non-covalent interactions between residues of protein binding domains and artificial surfaces.

The turnover of FXII-mediated catalytic networks on the platelet surface membrane is a potentially important mechanism in the response of hemostatic systems to foreign materials. We have shown that when normal platelets are freed of plasma proteins and recombined with FXIIa-deficient plasma, the response of the platelets to PGlcNac nanofibers is still sensitive to corn trypsin inhibitor [64]. This indicates that a platelet-bound pool of FXII is important in the activation of the intrinsic coagulation cascade, and is consistent with findings [65] that the platelet surface glycoprotein GP1b (the von Willebrand factor receptor or CD42b) can bind FXII at an extracellular domain that might overlap with the GP1bα thrombin binding sites between residues 251 and 284 on the glycoprotein [66]. Based on platelet binding competition and biosensor studies, FXII and high MW kinin might bind to the same site on GP1bα65. This observation assumes additional importance in view of findings that high MW kinin disrupts the interaction between GP1b on platelets and the β-integrin Mac-1 on leukocytes [67]. Interestingly, high MW kininogen has been shown to inhibit fibrinogen binding to platelets [68], an effect reminiscent of the kininogen/fibrinogen "conversion" effect on glass [33]. These findings suggest that the vWf receptor might play a scaffolding role on the platelet surface for the organization of elements of the intrinsic coagulation and/or kallikrein/kinin systems.

The concerted function of FXII- and fibrinogen/integrin-mediated events in the hemostatic response to foreign materials is becoming appreciated. As will be discussed in more detail in the following section, FXIIa and integrin functions can be synergistic for platelet activation and intrinsic coagulation pathway turnover when platelets contact some

[64], but not all [63], artificial materials. What is not known is how the two systems couple. Does FXIIa affect processes through which integrins respond to surface-bound fibrinogen? Alternatively, do the two processes converge downstream of integrin activation, after integrin outside-in signaling generates an intracellular calcium signal for surface exposure of phosphatidylserine?

The adsorption of plasma proteins and platelets to foreign materials sets the stage for longer term-tissue responses to the material interface. The reaction of hemostatic systems to the artificial surface sets the stage for longer term responses involving both innate and acquired immunity as well as wound healing systems. Implantation of a foreign material can involve tissue trauma, an attendant inflammatory response, and a sequential wound healing reaction involving angiogenesis, fibroblast activation and matrix remodeling [69]. There is thus a ternary interaction between blood components, the artificial surface and injured tissue. For example, thrombin, generated at the material interface, can have an activating action on inflammatory processes. Many cell types have been shown to have PAR receptors [70-81], including macrophages [82,83], which are activated when exposed to thrombin for the generation of proinflammatory cytokines, chemokines, reactive oxygen species and metalloproteases [84-86]. Of particular importance is the activation of astrocytes in neuronal tissue by the action of thrombin on PAR [87-95] for tissue damage [93,96-102]. These observations concerning macrophage activation in neural tissue by thrombin have given rise to the general idea that this protease is toxic to neural tissues [103]. An important negative feedback system that could be operant at the tissue/foreign material interface involves thrombin activation of protein C. In addition to the well-understood mechanisms for inhibition of humoral coagulation, protein C exerts an anti-inflammatory effect (see Esmon for a review) [104] on a wide variety of tissues. These target tissues include endothelial cells, where inhibition occurs in a wide variety of NF-κB-mediated functions, notably adhesion molecule expression and EC retraction. Protein C can also affect monocyte/macrophages via down regulation of tissue factor expression in response to Th1 cytokines [105]. Another negative feedback system that counteracts humoral coagulation cascade turnover involves factor XIIa (Hageman factor) activation [106] of kallikrein/bradykinin systems for thrombolysis [107]. The information at hand indicates that the contact activation branch of the coagulation cascade is of lesser importance than bradykinin-mediated thrombolysis (see Agostoni for a review) [108]. As of this writing, the net inflammatory effect of local thrombin generation at artificial interfaces is poorly understood.

Following blood component arrival at the injury/implantation site, platelets come into contact with exposed collagen and other aspects of the extracellular matrix at the edges of the foreign material. This contact of platelet with both the artificial surface and the extracellular matrix contact stimulates the platelets to release clotting factors, and growth factors, *e.g.*, platelet-derived growth factor (PDGF) and transforming growth factor beta (TGF-β). Once hemostasis has been achieved, neutrophils can enter the wound site and begin to remove bacteria and damaged tissue. Subsequently, macrophages arrive and release additional PDGF and TGF-β. Fibroblasts can now migrate into the "clean" wound site, proliferate and deposit new extracellular matrix. Enzymatic cross-linking of the collagenous matrix initiates the remodeling phase. Numerous normal cell-cell and cell-matrix signaling events are required for proper consolidation of the new foreign material connective tissue interface [109].

The effect of artificial surface structure on hemostatic system activation is an important consideration in the knowledge-based design of medical devices. Altering surface structure to maximize this activating interaction is desirable for topical hemostasis products. Minimizing hemostatic activation is expected to result in a more passive surface, an attribute important for many blood-contacting products, including stents, prosthetic heart valves, parts of extracorporealization devices, and containers for *in vitro* blood and blood component storage. Consideration of the hemostatic effects of foreign materials is anticipated to become even more important as nanotechnologies develop because of the vastly increased specific surface areas of nanometer-scale structures.

The Interaction of Hemostatic Systems with Poly-*N*-acetyl glucosamine Nanofibers: An Example of the Three-Step Interaction Mechanism

In this section the concerted function of the plasma protein adsorption, conformational distortion and hemostatic system activation steps will be examined as they occur on the surface of poly-*N*-acetyl glucosamine (pGlcNAc) nanofibers. This new material is used in products to control bleeding from injured tissue surfaces in trauma and surgery. Until 1999, when the a pGlcNAc-based product (the Syvek™ patch, Marine Polymer Technologies, Danvers, MA) was approved by the United States Food and Drug Administration for hemorrhage control at vascular access sites after catheterization procedures, the standard product for surface hemostasis was cotton-cellulosic gauze. The Syvek™ patch and several additional approved hemostatic products are composed of ultrapure pGlcNAc nanofibers produced in large-scale cGMP cultures of a marine microalga/diatom (Vournakis *et al.*) [110,111]. The three-step interaction mechanism for hemostasis has been extensively investigated with the pGlcNAc nanofibers.

Protein Adsorption to pGlcNAc-Nanofibers

The composition of the pGlcNAc plasma protein adsorption proteome is a specific consequence of the primary, ternary and quaternary structure of this new biomaterial. The structural properties of pGlcNAc nanofibers have been studied using a variety of methods, including NMR, circular dichroism, FTIR, electron microscopy, x-ray diffraction, unidirectional chitinase degradation, and microdiffraction electron crystallography [66,67,79,80,81,82]. It has been shown by the set of structural and enzymatic studies in the prior several references that the fibrils formed by the diatom have a unique tertiary structure referred to as the beta-configuration. In this crystallographic form, the individual polymer molecules are lined parallel to one another with the reducing end of each polymer at one end of the fiber, and the non-reducing ends at the other end. The dimensions of the fibrils have been reported to be [79] 60-80μ in length, 100-200nm in width, and ~50nm in thickness. It is estimated that there are between 350-2000 parallel poly-*N*-acetyl glucosamine chains per fibril. NMR and FTIR studies [66,67] have shown that the molecular orientations of the individual polymer chains in the nanofibers are limited by specific inter-chain hydrogen-bonding (see Figure 1). The resultant nanofiber therefore has a structural integrity that is both unique, and cannot be disrupted by normal physiological conditions, but only by highly chaotropic solvents.

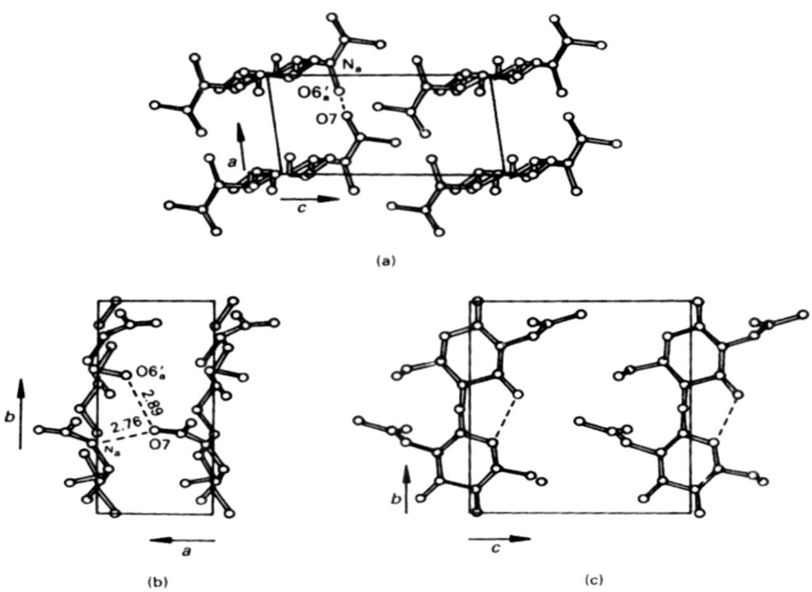

Figure 1- Hydrogen Bonding Tertiary Structure pGlcNAc Nanofibers

The incubation of pGlcNAc nanofibers with plasma results in the selective adsorption of a specific set of plasma proteins. To investigate this phenomena plasma was incubated with pGlcNAc then the fibers were centrifugally washed as detailed elsewhere [64]. pGlcNAc-bound proteins were labeled with the red-fluorescent dye Cy5-NHS ester while total plasma proteins were covalently modified with a green-fluorescent dye Cy3-NHS ester (Amersham, Inc. Piscataway, NJ). The dye conjugates preserved the charge on modified amino acid side chains so as not to interfere with the isoelectric focusing step. Equal mass mixtures of total plasma and material adsorbed proteins were co-electrophoresed, and the fluorescent color (more red fluorescent for proteins that are selectively adsorbed, green fluorescent for excluded proteins, and yellow for intermediate cases) is used to judge interaction selectivity (see Figure 2). The data in Figure 2 below show that many plasma proteins (green in the middle frame, some having been identified with GC/mass spectrometry analysis) were selectively adsorbed to the material. Other proteins were excluded (orange in the middle frame).

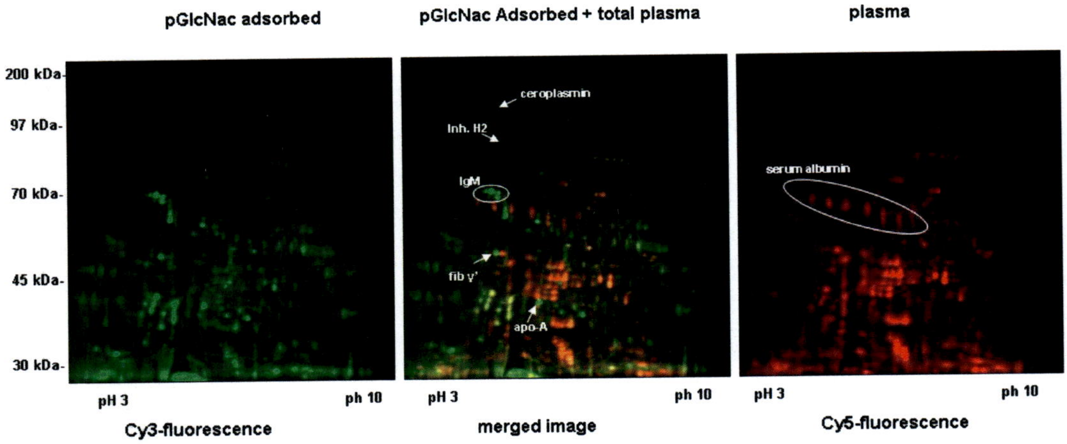

Figure 2- 2D Isoelectric Focusing/SDS Polyacrylamide Gel Electrophoresis Differential Analysis of pGlcNAc Adsorbed and Total Plasma Proteins

A comparison of the set of plasma proteins (adsorption proteome) that bind to pGlcNAc as compared to other polyglucosamine-containing materials, such as chitin, chitosan and a chitosan-based product Chito-Seal^{TM} (Abbott Vascular, Inc., Redwood City, CA), is shown in Figure 3. This study demonstated that pGlcNAc bound the most and greatest number of proteins, including the 67 kDa serum albumin, 70 kDa IgM heavy (μ) chain and 50 – 60 kDa fibrinogen chains. Chitosan and Chito-Seal^{TM} selectively bound 120 kDa ceroplasmin (ferroxidase) and 107 kDa inter-α globin inhibitor H2, while β-pGlcNAc fibers selectively bound apoliprotein. There are significant differences between the profiles of plasma proteins that tightly adsorb to different materials. The pGlcNAc nanofibers specifically adsorbs Apo A2, while chitin, chitosan and Chito-Seal^{TM} bind inter-α globin inhibitor H2 and ceroplasmin. All the materials bind fibrinogen, serum albumin, and IgM to a greater or lesser extent. The conformation of the adsorbed fibrinogen (and fibrin chains after clot formation) is potentially different for each material, and may be a factor in explaining the specificity of the interaction for platelet activation (see Fischer *et al.* for greater detail) [64]. The tendency of IgM to adsorb strongly to all of the tested materials is of potential importance with respect to complement activation and the relationship of artificial surfaces to innate and acquired immune systems.

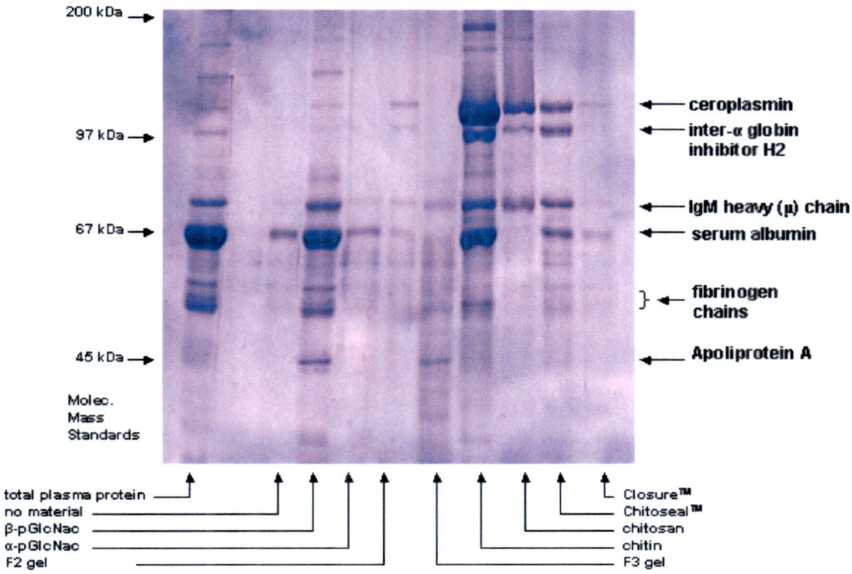

Figure 3- Differential Plasma Protein Binding to Glucosamine-based Materials

Hemostatic System Activation in Response to Plasma Protein Adsorption to pGlcNAc

The initial adsorption of plasma proteins to pGlcNAc after the marine nanofiber contacts blood is probably largely completed on a millisecond timescale. In the ensuing seconds and minutes the intrinsic coagulation cascade turns over for thrombin generation and platelets are activated [112]. The contact of platelets with pGlcNAc has been shown to result in a platelet activation response that includes shape change, pseudopodia extension [112], intracellular calcium signal generation, p-selectin and phosphatidyl serine (PS) surface exposure, integrin activation [113] and factor X binding [114]. The kinetics of FXIIa activation for intrinsic coagulation cascade turnover and platelet activation were substantially accelerated on pGlcNac fibers as compared to other materials in use for surface hemostasis [112].

Several observations indicate that the platelet activation response to pGlcNAc is dependent on both FXII and fibrinogen adsorption to the marine nanofiber. Platelet activation by pGlcNAc was antagonized by inhibition of both FXIIa, with corn trypsin inhibitor, and integrins, with eptifibatide (see Fischer *et al*for details) [64]. Experiments with biotinylated platelets show that a specific subset of platelet surface proteins, including the von Willebrand factor receptor glycoprotein Ib and the fibrinogen receptor integrin $\alpha_{IIb}\beta_3$, tightly associate with pGlcNAc fibers. The specific subset of platelet surface proteins that bind to pGlcNAc was not altered when integrin $\alpha_{IIb}\beta_3$ complex, thrombin or factor XIIa was inhibited (respectively with eptifibatide, low molecular weight heparin or corn trypsin inhibitor). This result suggests that there are multiple interaction points between surface proteins and the marine polymer fibrils resulting in a high-affinity that is not dependent on activation-related conformations of the proteins. We do not know why only a specific subset of platelet proteins bind tightly to pGlcNAc, but the finding that the interaction of red blood cells with pGlcNAc is inhibited by neuraminidase activity [113] indicates that negatively charged surface carbohydrates might be involved. These findings are consistent with the hypothesis that FXII and fibrinogen are conformationally altered when adsorbed to pGlcNAc.

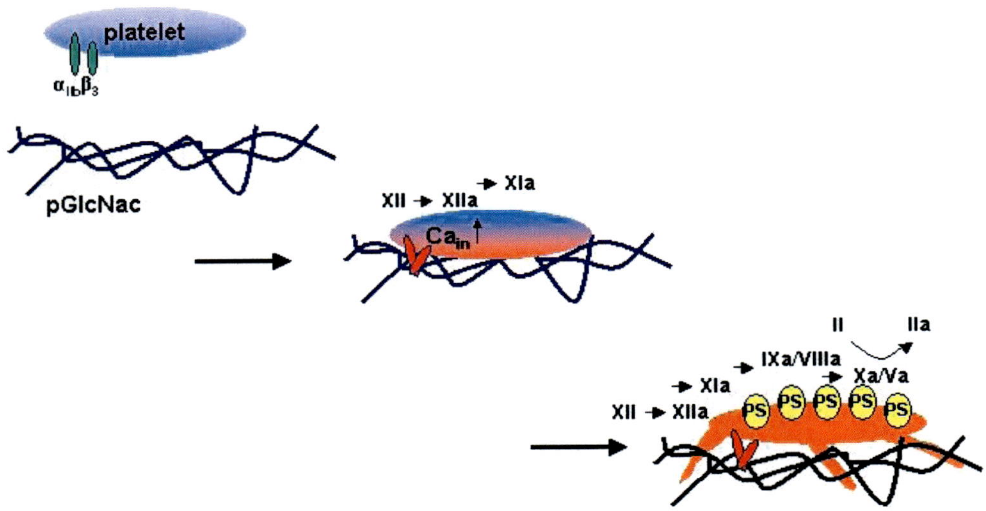

Figure 4- Mechanism of Platelet Interaction With pGlcNAc for Thrombin Generation

A four-step process can be hypothesized to explain how pGlcNAc/platelet acceleration of fibrin gel formation takes place (see Figure 4). First, plasma proteins bind to pGlcNAc in a tight, but nonspecific manner through chemical/physical adsorption processes that involve essentially all the major plasma proteins. Secondly, a complex subset of surface-platelet proteins mediates the attachment of the cell to the pGlcNAc/plasma protein matrix. The third step is an activation of integrin outside-in signaling. The final step in pGlcNAc/platelet mediation of fibrin polymerization is an acceleration of the intrinsic (contact) coagulation pathway for thrombin generation on the platelet (PS rich) interface. The observation that eptifibatide removes the ability of platelet-pGlcNAc mixtures to accelerate fibrin polymerization indicates that factor XIIa function (on the level of the platelet) is dependent on integrin activation, perhaps *due to the coupling between integrin signaling and PS surface exposure.* In conclusion, studies with pGlcNAc define a set of mechanistic processes that demonstrate the protein adsorption, conformational distortion and system activation three-step process.

Pre-clinical and Clinical Evaluation of pGlcNAc-based Products

In pre-clinical and clinical practice pGlcNAc-mediated platelet and intrinsic coagulation activation mechanisms for hemorrhage control are augmented in a redundant manner by two other hemostatic processes. First, red blood cell stimulation occurs upon contact with the marine polymer [115,116] and releases prostaglandins and serotonin. Secondly, pGlcNAc has been shown to mediate vasoconstriction through mechanisms that are endothelium-dependent and in part mediated by endothelin-1 [117]. The redundancy of platelet-activation, red blood cell agglutination and vasoactive mechanisms for hemostasis by pGlcNAc provides an explanation for the ability of this polymeric material to provide hemostasis in a wide variety of animal model systems as well as in human clinical use.

Systematic clinical evaluations and comparisons of the different products for surface hemostasis have not been performed. However, gauze, two chitosan-based products (Chito-SealTM, Clo-SurTM) and a pGlcNAc-based product (SyvekTM patch) have been compared in a porcine splenic injury model [112]. These investigators measured the number of one-minute compression cycles required to achieve bleeding cessation from splenic injury sites. Hemorrhage was stopped after three minutes with the SyvekTM patch, while Chito-SealTM and Clo-SurTM and gauze products required eight to ten minutes or longer. Three other porcine spleen studies [118-120] demonstrated that the pGlcNAc-based SyvekTM patch was superior to other hemostatic products such as absorbable collagen (ActifoamTM), fibrin sealant (BiohealTM), fibrin bandage (TachocombTM), fibrin glue (TisseelTM), and oxidized regenerated cellulose (SurgicelTM) for providing surface hemostasis.

The pGlcNAc-based products have also been shown to be efficacious in a human clinical study in colon surgery patients[121]. The SyvekTM patch has been compared with gauze in a human clinical study on cardiac catheterization in which the endpoint was bleeding cessation after femoral catheter withdrawal [122]. In this study, holding pressure following catheter removal was carefully controlled by a pressure regulation device. The pGlcNAc-based SyvekTM patch was shown to reduce the time for bleeding cessation by 40% as compared to gauze. The ability of the pGlcNAc-based product to accelerate hemostasis as compared to gauze in these two studies correlates with the *in vitro* observations reviewed above that pGlcNAc fibers are the most pro-hemostatic with respect to activation of platelets and humeral coagulation systems [63,112].

A pGlcNAc-based product, the mRDH bandage, has recently been given FDA clearance "intended for the temporary control of severely bleeding wounds." A prospective observational clinical study in humans examined the efficacy of the mRDH in patients with hepatic injury who had failed all other available methods of hemostasis.[83] The mRDH dressing was used in a real-life emergency situation, in patients with Grade V (n=2), Grade IV (n=5), or Grade III (n=3) injury, with no application of the Pringle maneuver or "complex surgical procedures." The success of the mRDH in controlling bleeding in these patients was impressive. Complete cessation of bleeding was noted in 9/10 patients within 5 minutes, including one patient with iliac vein laceration.123 One patient died because a retrohepatic vein laceration was missed. [83] The survival rate of 90% is remarkable considering the expected mortality of hepatic injuries of Grade III to V (24–80%) [84,85]. This study [83] supports previously obtained animal data[86] and provides evidence that mRDH is efficacious in controlling low pressure/high flow (venous) bleeding.

The mRDH trauma product successfully prevented death in the majority of cases where exsanguination from abdominal wounds was certain in the clinical study noted above [83]. In addition, a recently initiated clinical registry study has shown that the mRDH is effective in controlling bleeding on more than 90 patients with a wide variety of trauma related wounds. It was reasonable to predict, based on the ability of pGlcNAc fibers to adsorb plasma proteins for activation of hemostatic systems, that pGlcNAc-based products would perform well in actual use for surface hemostasis. The preponderance of human and animal *in vivo* data with pGlcNAc-based products has born out this prediction.

Conclusion

This review has outlined basic principles that have been learned from over fifty years of analysis of the interactions of a few materials with hemostatic systems. These principles are now being applied for the rational design of medical products. The response of hemostatic systems to artificial surfaces is hypothesized to occur in three stages: An initial selective adsorption event that is a function of a material's surface structure and chemical properties; a conformational distortion of the adsorbed proteins; and a functional reaction that is a consequence of the conformational alterations of the adsorbed proteins. The interaction of hemostatic systems with pGlcNAc nanofibers is examined in detail to illustrate how the aforementioned three-step interaction process functions. Preclinical and clinical results with pGlcNAc demonstrate the utility of considering the three-step mechanism for the rational design of products for surface hemostasis.

References

[1] Packham MA. The behavior of platelets at foreign surfaces. *Proc Soc Exp Biol Med.* 1988;189:261-274

[2] Hippocrates. On the surgery. *http://www.skepticfiles.org/think/surgery.htm*. 400 BC

[3] Doone A. De L'origine des globules du sang, de leur mode de formation et de leur. *Fin. Compt. Rend. Sean. L'acad. Sci.* 1842;14:366-368

[4] Bizzozero G. Ueber einen neuen formbestandtheil des blutes und dessen rolle bie der thromboses and der blutgerinnung. Archiv. *Path. Anatom. Physiol. Klin. Med.* 1881;90:261-332

[5] Bizzoxzero G. Sur un nouvel element: morphologique du sang chez les mammiferes et sur son. *Arch Ital Biol.* 1881;1881:1-4

[6] Buchanan A. On the coagulation of blood. London Medical Gazette. 1845;1, New Series:617-621

[7] Virchow R. Besammelte Abhandlungen zur wissenschaftlichen Medicin. Thrombose und Embolie. Gefassentzundung und septishce iInfektion. *Frankfurt am Main: Von Meidinger and Sohn*; 1856:219-732

[8] Wright J, Minor G. The viscous metamorphosis of the blood platelets. *J. Exp. Med.* 1917;26:395-409

[9] Barr H. The stickiness of platelets. *Lancet.* 1941;238:609-610

[10] Ratnoff OD, Margolius A, Jr. Hageman trait: an asymptomatic disorder of blood coagulation. *Trans Assoc Am Physicians.* 1955;68:149-154

[11] Ratnoff OD, Rosenblum JM. Role of Hageman factor in the initiation of clotting by glass; evidence that glass frees Hageman factor from inhibition. *Am J Med.* 1958;25:160-168

[12] Stouffer JE, Lipscomb HS. Reversible binding of ACTH to a glass surface. *Endocrinology.* 1963;72:91-94

[13] Lissitzky S, Roques M, Benevent MT. [Influence of proteins on the adsorption of I 131-labelled L-thyroxin on glass.]. *C R Seances Soc Biol Fil.* 1960;154:396-399
[14] Bull HB. Adsorption of bovine serum albumin on glass. *Biochim Biophys Acta.* 1956;19:464-471
[15] Sit PS, Marchant RE. Surface-dependent conformations of human fibrinogen observed by atomic force microscopy under aqueous conditions. *Thromb Haemost.* 1999;82:1053-1060
[16] Rapoza RJ, Horbett TA. Postadsorptive transitions in fibrinogen: influence of polymer properties. *J Biomed Mater Res.* 1990;24:1263-1287
[17] Perez-Luna VH, Horbett TA, Ratner BD. Developing correlations between fibrinogen adsorption and surface properties using multivariate statistics. Student Research Award in the Doctoral Degree Candidate Category, 20th annual meeting of the Society for Biomaterials, Boston, MA, April 5-9, 1994. *J Biomed Mater Res.* 1994;28:1111-1126
[18] Silberberg A. The absorption of flexible macromolecules Part I: The isolated macromolecule at a plane interface. *J. Physical Chem.* 1962;66:1872-1883
[19] Vroman L, Adams AL. Possible involvement of fibrinogen and proteolysis in surface activation. A study with the recording ellipsometer. *Thromb Diath Haemorrh.* 1967;18:510-524
[20] Vroman L, Adams AL. Effect of heparin on reactions at aminated polymer--blood interfaces. *J Colloid Interface Sci.* 1969;31:188-195
[21] Zucker MB, Vroman L. Platelet adhesion induced by fibrinogen adsorbed onto glass. *Proc Soc Exp Biol Med.* 1969;131:318-320
[22] Scarborough DE, Mason RG, Dalldorf FG, Brinkhous KM. Morphologic manifestations of blood-solid interfacial reactions. A scanning and transmission electron microscopic study. *Lab Invest.* 1969;20:164-169
[23] Adams AL, Vroman L. Observatin of Factor Xii and Factor Xii Deficient Plasma at an Interface. *Thromb Diath Haemorrh.* 1965;13:387-393
[24] Newmark P, Patel N. The adsorption of intrinsic factor concentrate to glass: occurrence and prevention with reference to radioisotope dilution assays of vitamin B 12. *Blood.* 1971;38:524-528
[25] Chiu TH, Nyilas E, Lederman DM. Thermodynamics of native protein/foreign surface interactions. IV. Calorimetric and microelectrophoretic study of human fibrinogen sorption onto glass and LTI-carbon. *Trans Am Soc Artif Intern Organs.* 1976;22:498-513
[26] Lindon JN, McManama G, Kushner L, Merrill EW, Salzman EW. Does the conformation of adsorbed fibrinogen dictate platelet interactions with artificial surfaces? *Blood.* 1986;68:355-362
[27] Heitz F, Van Mau N. Protein structural changes induced by their uptake at interfaces. *Biochim Biophys Acta.* 2002;1597:1-11
[28] Liu G, Shen Y, Atreya HS, Parish D, Shao Y, Sukumaran DK, Xiao R, Yee A, Lemak A, Bhattacharya A, Acton TA, Arrowsmith CH, Montelione GT, Szyperski T. NMR data collection and analysis protocol for high-throughput protein structure determination. *Proc Natl Acad Sci U S A.* 2005;102:10487-10492
[29] Shen Y, Atreya HS, Liu G, Szyperski T. G-matrix Fourier transform NOESY-based protocol for high-quality protein structure determination. J *Am Chem Soc.* 2005;127:9085-9099

[30] Drobny GP, Long JR, Karlsson T, Shaw W, Popham J, Oyler N, Bower P, Stringer J, Gregory D, Mehta M, Stayton PS. Structural studies of biomaterials using double-quantum solid-state NMR spectroscopy. *Annu Rev Phys Chem.* 2003;54:531-571

[31] Long JR, Shaw WJ, Stayton PS, Drobny GP. Structure and dynamics of hydrated statherin on hydroxyapatite as determined by solid-state NMR. *Biochemistry.* 2001;40:15451-15455

[32] Read MJ, Burkett SL. Asymmetric alpha -helicity loss within a peptide adsorbed onto charged colloidal substrates. *J Colloid Interface Sci.* 2003;261:255-263

[33] Vroman L, Adams AL, Fischer GC, Munoz PC. Interaction of high molecular weight kininogen, factor XII, and fibrinogen in plasma at interfaces. *Blood.* 1980;55:156-159

[34] Coller BS, Peerschke EI, Scudder LE, Sullivan CA. A murine monoclonal antibody that completely blocks the binding of fibrinogen to platelets produces a thrombasthenic-like state in normal platelets and binds to glycoproteins IIb and/or IIIa. *J Clin Invest.* 1983;72:325-338

[35] Goodman SL, Cooper SL, Albrecht RM. Integrin receptors and platelet adhesion to synthetic surfaces. *J Biomed Mater Res.* 1993;27:683-695

[36] De Scheerder I, Verbeken E, Van Humbeeck J. Metallic surface modification. *Semin Interv Cardiol.* 1998;3:139-144

[37] Nagai H, Handa M, Kawai Y, Watanabe K, Ikeda Y. Evidence that plasma fibrinogen and platelet membrane GPIIb-IIIa are involved in the adhesion of platelets to an artificial surface exposed to plasma. *Thromb Res.* 1993;71:467-477

[38] Wang YX, Robertson JL, Spillman WB, Jr., Claus RO. Effects of the chemical structure and the surface properties of polymeric biomaterials on their biocompatibility. *Pharm Res.* 2004;21:1362-1373

[39] Hayward JA, Chapman D. Biomembrane surfaces as models for polymer design: the potential for haemocompatibility. *Biomaterials.* 1984;5:135-142

[40] Juliano RL. Signal transduction by cell adhesion receptors and the cytoskeleton: functions of integrins, cadherins, selectins, and immunoglobulin-superfamily members. *Annu Rev Pharmacol Toxicol.* 2002;42:283-323

[41] Baglia FA, Walsh PN. Thrombin-mediated feedback activation of factor XI on the activated platelet surface is preferred over contact activation by factor XIIa or factor XIa. *J Biol Chem.* 2000;275:20514-20519

[42] Colman RW, Schmaier AH. Contact system: a vascular biology modulator with anticoagulant, profibrinolytic, antiadhesive, and proinflammatory attributes. *Blood.* 1997;90:3819-3843

[43] Renne T, Pozgajova M, Gruner S, Schuh K, Pauer HU, Burfeind P, Gailani D, Nieswandt B. Defective thrombus formation in mice lacking coagulation factor XII. *J Exp Med.* 2005;202:271-281

[44] Clarke BJ, Cote HC, Cool DE, Clark-Lewis I, Saito H, Pixley RA, Colman RW, MacGillivray RT. Mapping of a putative surface-binding site of human coagulation factor XII. *J Biol Chem.* 1989;264:11497-11502

[45] Citarella F, Ravon DM, Pascucci B, Felici A, Fantoni A, Hack CE. Structure/function analysis of human factor XII using recombinant deletion mutants. Evidence for an additional region involved in the binding to negatively charged surfaces. *Eur J Biochem.* 1996;238:240-249

[46] Nuijens JH, Huijbregts CC, Eerenberg-Belmer AJ, Meijers JC, Bouma BN, Hack CE. Activation of the contact system of coagulation by a monoclonal antibody directed against a neodeterminant in the heavy chain region of human coagulation factor XII (Hageman factor). *J Biol Chem*. 1989;264:12941-12949
[47] Fair BD, Saito H, Ratnoff OD, Rippon WB. Detection by fluorescence of structural changes accompanying the activation of Hageman factor (factor XII). *Proc Soc Exp Biol Med*. 1977;155:199-202
[48] Donaldson VH, Ratnoff OD. Hageman factor: alterations in physical properties during activation. *Science*. 1965;150:754-756
[49] Cochrane CG, Revak SD, Wuepper KD. Activation of Hageman factor in solid and fluid phases. A critical role of kallikrein. *J Exp Med*. 1973;138:1564-1583
[50] Revak SD, Cochrane CG, Johnston AR, Hugli TE. Structural changes accompanying enzymatic activation of human Hageman factor. *J Clin Invest*. 1974;54:619-627
[51] Revak SD, Cochrane CG, Griffin JH. The binding and cleavage characteristics of human Hageman factor during contact activation. A comparison of normal plasma with plasmas deficient in factor XI, prekallikrein, or high molecular weight kininogen. *J Clin Invest*. 1977;59:1167-1175
[52] Revak SD, Cochrane CG, Bouma BN, Griffin JH. Surface and fluid phase activities of two forms of activated Hageman factor produced during contact activation of plasma. *J Exp Med*. 1978;147:719-729
[53] Espana F, Ratnoff OD. The role of prekallikrein and high-molecular-weight kininogen in the contact activation of Hageman factor (factor XII) by sulfatides and other agents. *J Lab Clin Med*. 1983;102:487-499
[54] Espana F, Ratnoff OD. Activation of Hageman factor (factor XII) by sulfatides and other agents in the absence of plasma proteases. J Lab Clin Med. 1983;102:31-45
[55] Cochrane CG, Griffin JH. Molecular assembly in the contact phase of the Hageman factor system. *Am J Med*. 1979;67:657-664
[56] Henry ML, Everson B, Ratnoff OD. Inhibition of the activation of Hageman factor (factor XII) by beta 2-glycoprotein I. *J Lab Clin Med*. 1988;111:519-523
[57] Dumenco LL, Everson B, Culp LA, Ratnoff OD. Inhibition of the activation of Hageman factor (factor XII) by platelet factor 4. *J Lab Clin Med*. 1988;112:394-400
[58] Rehmus EH, Greene BM, Everson BA, Ratnoff OD. Inhibition of the activation of Hageman factor (factor XII) by complement subcomponent C1q. *J Clin Invest*. 1987;80:516-521
[59] Zuraw BL, Curd JG. Demonstration of modified inactive first component of complement (C1) inhibitor in the plasmas of C1 inhibitor-deficient patients. *J Clin Invest*. 1986;78:567-575
[60] Revak SD, Cochrane CG. Hageman factor: its structure and modes of activation. *Thromb Haemost*. 1976;35:570-575
[61] Griffin JH. Role of surface in surface-dependent activation of Hageman factor (blood coagulation factor XII). *Proc Natl Acad Sci U S A*. 1978;75:1998-2002
[62] Hojima Y, Cochrane CG, Wiggins RC, Austen KF, Stevens RL. In vitro activation of the contact (Hageman factor) system of plasma by heparin and chondroitin sulfate *E*. *Blood*. 1984;63:1453-1459
[63] Fischer TH, Bode, A., Demcheva, M. and Vournakis, J. Hemostatic properties of glucosamine-based materials. *Glycobiology*. 2005;submitted

[64] Fischer TH, Thatte HS, Nichols TC, Bender-Neal DE, Bellinger AD, Vournakis JN. Synergistic platelet integrin signaling and factor XII activation in poly-N-acetyl glucosamine fiber-mediated hemostasis. *Biomaterials*. 2005;26:5433-5443

[65] Bradford HN, Pixley RA, Colman RW. Human factor XII binding to the glycoprotein Ib-IX-V complex inhibits thrombin-induced platelet aggregation. J *Biol Chem*. 2000;275:22756-22763

[66] Marchese P, Murata M, Mazzucato M, Pradella P, De Marco L, Ware J, Ruggeri ZM. Identification of three tyrosine residues of glycoprotein Ib alpha with distinct roles in von Willebrand factor and alpha-thrombin binding. *J Biol Chem*. 1995;270:9571-9578

[67] Chavakis T, Santoso S, Clemetson KJ, Sachs UJ, Isordia-Salas I, Pixley RA, Nawroth PP, Colman RW, Preissner KT. High molecular weight kininogen regulates platelet-leukocyte interactions by bridging Mac-1 and glycoprotein Ib. *J Biol Chem*. 2003;278:45375-45381

[68] Gustafson EJ, Lukasiewicz H, Wachtfogel YT, Norton KJ, Schmaier AH, Niewiarowski S, Colman RW. High molecular weight kininogen inhibits fibrinogen binding to cytoadhesins of neutrophils and platelets. *J Cell Biol*. 1989;109:377-387

[69] Kirkpatrick CJ, Krump-Konvalinkova V, Unger RE, Bittinger F, Otto M, Peters K. Tissue response and biomaterial integration: the efficacy of in vitro methods. *Biomol Eng*. 2002;19:211-217

[70] Vergnolle N, Cellars L, Mencarelli A, Rizzo G, Swaminathan S, Beck P, Steinhoff M, Andrade-Gordon P, Bunnett NW, Hollenberg MD, Wallace JL, Cirino G, Fiorucci S. A role for proteinase-activated receptor-1 in inflammatory bowel diseases. *J Clin Invest*. 2004;114:1444-1456

[71] Marutsuka K, Hatakeyama K, Sato Y, Yamashita A, Sumiyoshi A, Asada Y. Protease-activated receptor 2 (PAR2) mediates vascular smooth muscle cell migration induced by tissue factor/factor VIIa complex. *Thromb Res*. 2002;107:271-276

[72] Kawabata A, Kuroda R, Nishida M, Nagata N, Sakaguchi Y, Kawao N, Nishikawa H, Arizono N, Kawai K. Protease-activated receptor-2 (PAR-2) in the pancreas and parotid gland: Immunolocalization and involvement of nitric oxide in the evoked amylase secretion. *Life Sci*. 2002;71:2435-2446

[73] Iwakiri K, Ghazizadeh M, Jin E, Fujiwara M, Takemura T, Takezaki S, Kawana S, Yasuoka S, Kawanami O. Human airway trypsin-like protease induces PAR-2-mediated IL-8 release in psoriasis vulgaris. *J Invest Dermatol*. 2004;122:937-944

[74] Knight DA, Lim S, Scaffidi AK, Roche N, Chung KF, Stewart GA, Thompson PJ. Protease-activated receptors in human airways: upregulation of PAR-2 in respiratory epithelium from patients with asthma. *J Allergy Clin Immuno*l. 2001;108:797-803

[75] Steinhoff M, Corvera CU, Thoma MS, Kong W, McAlpine BE, Caughey GH, Ansel JC, Bunnett NW. Proteinase-activated receptor-2 in human skin: tissue distribution and activation of keratinocytes by mast cell tryptase. *Exp Dermatol*. 1999;8:282-294

[76] Bohm H, Brinkmann V, Drab M, Henske A, Kurzchalia TV. Mammalian homologues of C. elegans PAR-1 are asymmetrically localized in epithelial cells and may influence their polarity. *Curr Biol*. 1997;7:603-606

[77] Brass LF, Molino M. Protease-activated G protein-coupled receptors on human platelets and endothelial cells. *Thromb Haemost*. 1997;78:234-241

[78] Wang J, Zheng H, Hollenberg MD, Wijesuriya SJ, Ou X, Hauer-Jensen M. Up-regulation and activation of proteinase-activated receptor 2 in early and delayed

radiation injury in the rat intestine: influence of biological activators of proteinase-activated receptor 2. *Radiat Res*. 2003;160:524-535
[79] Liu J, Bastian M, Kohlschein P, Schuff-Werner P, Steiner M. Expression of functional protease-activated receptor 1 in human prostate cancer cell lines. *Urol Res*. 2003;31:163-168
[80] Toyoda N, Gabazza EC, Inoue H, Araki K, Nakashima S, Oka S, Taguchi Y, Nakamura M, Suzuki Y, Taguchi O, Imoto I, Suzuki K, Adachi Y. Expression and cytoprotective effect of protease-activated receptor-1 in gastric epithelial cells. *Scand J Gastroenterol*. 2003;38:253-259
[81] Faraut B, Barbier J, Ravel-Chapuis A, Doyennette MA, Jandrot-Perrus M, Verdiere-Sahuque M, Schaeffer L, Koenig J, Hantai D. Thrombin downregulates muscle acetylcholine receptors via an IP3 signaling pathway by activating its G-protein-coupled protease-activated receptor-1. *J Cell Physiol*. 2003;196:105-112
[82] Johansson U, Lawson C, Dabare M, Syndercombe-Court D, Newland AC, Howells GL, Macey MG. Human peripheral blood monocytes express protease receptor-2 and respond to receptor activation by production of IL-6, IL-8, and IL-1{beta}. J *Leukoc Biol*. 2005
[83] Fields RC, Schoenecker JG, Hart JP, Hoffman MR, Pizzo SV, Lawson JH. Protease-activated receptor-2 signaling triggers dendritic cell development. *Am J Pathol*. 2003;162:1817-1822
[84] Raza SL, Nehring LC, Shapiro SD, Cornelius LA. Proteinase-activated receptor-1 regulation of macrophage elastase (MMP-12) secretion by serine proteinases. *J Biol Chem*. 2000;275:41243-41250
[85] Szaba FM, Smiley ST. Roles for thrombin and fibrin(ogen) in cytokine/chemokine production and macrophage adhesion in vivo. *Blood*. 2002;99:1053-1059
[86] Bar-Shavit R, Kahn A, Wilner GD, Fenton JW, 2nd. Monocyte chemotaxis: stimulation by specific exosite region in thrombin. *Science*. 1983;220:728-731
[87] Perraud F, Besnard F, Sensenbrenner M, Labourdette G. Thrombin is a potent mitogen for rat astroblasts but not for oligodendroblasts and neuroblasts in primary culture. *Int J Dev Neurosci*. 1987;5:181-188
[88] Nelson RB, Siman R. Thrombin and its inhibitors regulate morphological and biochemical differentiation of astrocytes in vitro. *Brain Res Dev Brain Res*. 1990;54:93-104
[89] Cunningham DD. Regulation of neuronal cells and astrocytes by protease nexin-1 and thrombin. *Ann N Y Acad Sci*. 1992;674:228-236
[90] Beecher KL, Andersen TT, Fenton JW, 2nd, Festoff BW. Thrombin receptor peptides induce shape change in neonatal murine astrocytes in culture. *J Neurosci Res*. 1994;37:108-115
[91] Grabham P, Cunningham DD. Thrombin receptor activation stimulates astrocyte proliferation and reversal of stellation by distinct pathways: involvement of tyrosine phosphorylation. *J Neurochem*. 1995;64:583-591
[92] Bhat NR, Zhang P, Hogan EL. Thrombin activates mitogen-activated protein kinase in primary astrocyte cultures. *J Cell Physiol*. 1995;165:417-424
[93] Suidan HS, Niclou SP, Monard D. The thrombin receptor in the nervous system. *Semin Thromb Hemost*. 1996;22:125-133

[94] Wang H, Ubl JJ, Stricker R, Reiser G. Thrombin (PAR-1)-induced proliferation in astrocytes via MAPK involves multiple signaling pathways. *Am J Physiol Cell Physiol.* 2002;283:C1351-1364

[95] Donovan FM, Cunningham DD. Signaling pathways involved in thrombin-induced cell protection. *J Biol Chem*. 1998;273:12746-12752

[96] Nishino A, Suzuki M, Yoshimoto T, Otani H, Nagura H. A novel aspect of thrombin in the tissue reaction following central nervous system injury. *Acta Neurochir Suppl (Wien)*. 1994;60:86-88

[97] Donovan FM, Pike CJ, Cotman CW, Cunningham DD. Thrombin induces apoptosis in cultured neurons and astrocytes via a pathway requiring tyrosine kinase and RhoA activities. *J Neurosci*. 1997;17:5316-5326

[98] Kubo Y, Suzuki M, Kudo A, Yoshida K, Suzuki T, Ogasawara K, Ogawa A, Kurose A, Sawai T. Thrombin inhibitor ameliorates secondary damage in rat brain injury: suppression of inflammatory cells and vimentin-positive astrocytes. J *Neurotrauma. 2000*;17:163-172

[99] Xi G, Reiser G, Keep RF. The role of thrombin and thrombin receptors in ischemic, hemorrhagic and traumatic brain injury: deleterious or protective? *J Neurochem. 2003;84:3-9*

[100] Xue M, Del Bigio MR. Acute tissue damage after injections of thrombin and plasmin into rat striatum. *Stroke. 2001*;32:2164-2169

[101] Masada T, Xi G, Hua Y, Keep RF. The effects of thrombin preconditioning on focal cerebral ischemia in rats. *Brain Res*. 2000;867:173-179

[102] Hua Y, Wu J, Keep RF, Hoff JT, Xi G. Thrombin exacerbates brain edema in focal cerebral ischemia. *Acta Neurochir Supp*l. 2003;86:163-166

[103] Lundblad RL, Bradshaw RA, Gabriel D, Ortel TL, Lawson J, Mann KG. A review of the therapeutic uses of thrombin. *Thromb Haemost*. 2004;91:851-860

[104] Esmon CT, Fukudome K, Mather T, Bode W, Regan LM, Stearns-Kurosawa DJ, Kurosawa S. Inflammation, sepsis, and coagulation. *Haematologica*. Vol. 84; 1999:254-259

[105] Grey ST, Tsuchida A, Hau H, Orthner CL, Salem HH, Hancock WW. Selective inhibitory effects of the anticoagulant activated protein C on the responses of human mononuclear phagocytes to LPS, IFN-gamma, or phorbol ester. *J Immunol*. Vol. 153; 1994:3664-3672

[106] Kaplan AP, Joseph K, Silverberg M. Pathways for bradykinin formation and inflammatory disease. *J Allergy Clin Immunol*. Vol. 109; 2002:195-209

[107] Kluft C, Dooijewaard G, Emeis JJ. Role of the contact system in fibrinolysis. *Semin Thromb Hemost*. Vol. 13; 1987:50-68

[108] Agostoni A, Cugno M. [The kinin system: biological mechanisms and clinical implications]. *Recenti Prog Med*. Vol. 92; 2001:764-773

[109] Diegelmann RF, Evans MC. Wound healing: an overview of acute, fibrotic and delayed healing. *Front Biosci*. 2004;9:283-289

[110] Vournakis JN, Demcheva M, Whitson A, Guirca R, Pariser ER. Isolation, purification, and characterization of poly-N-acetyl glucosamine use as a hemostatic agent. *J Trauma*. 2004;57:S2-6

[111] Vournakis J, Finkielsztein, S., Pariser, E., Helton, M. Methods and composition s for poly-beta-4-N-acetylglucosamine biological barriers. United States of America; 1997:US patent 5,624,679

[112] Fischer TH, Connolly R, Thatte HS, Schwaitzberg SS. Comparison of structural and hemostatic properties of the poly-N-acetyl glucosamine Syvek Patch with products containing chitosan. *Microsc Res Tech*. 2004;63:168-174

[113] Thatte HS, Zagarins S, Khuri SF, Fischer TH. Mechanisms of poly-N-acetyl glucosamine polymer-mediated hemostasis: platelet interactions. *J Trauma*. 2004;57:S13-21

[114] Valeri CR, Srey R, Tilahun D, Ragno G. In vitro effects of poly-N-acetyl glucosamine on the activation of platelets in platelet-rich plasma with and without red blood cells. *J Trauma*. 2004;57:S22-25; discussion S25

[115] Thatte H, Zagarins S, Amiji M, Khuri S. Mechanisms of poly-N-acetylglucosamine mediated hemostasis: Red blood cell interactions. *J. Trauma-Inj. Inf. Crit. Care*. Vol. in press; 2003

[116] Valeri R. In vitro effects of poly-N-acetyl glucosamine on platelet rich plasma with and without RBCs on the activation of platelets and RBCs leading to hemostasis. *J. Trauma-Inj. Inf. Crit. Care*. Vol. in press; 2003

[117] Ikeda Y, Young LH, Vournakis JN, Lefer AM. Vascular effects of poly-N-acetylglucosamine in isolated rat aortic rings. *J Surg Res*. 2002;102:215-220

[118] Chan MW, Schwaitzberg SD, Demcheva M, Vournakis J, Finkielsztein S, Connolly RJ. Comparison of poly-N-acetyl glucosamine (P-GlcNAc) with absorbable collagen (Actifoam), and fibrin sealant (Bolheal) for achieving hemostasis in a swine model of splenic hemorrhage. *J Trauma*. 2000;48:454-457; discussion 457-458

[119] Schwaitzberg SD, Chan MW, Connolly RJ. Comparison of poly-N-acetyl glucosamine with absorbant collagen, oxidized regenerated celluose and chitosan for achieving hemostasis in coagulopathic animal models of splenic hemorrhage. *J. Trauma-Inj. Inf. Crit. Care*. 2003;in press

[120] Schwaitzberg SD, Chan MW, Cole DJ, Read M, Nichols T, Bellinger D, Connolly RJ. Comparison of poly-N-acetyl glucosamine with commercially available topical hemostats for achieving hemostasis in coagulopathic models of splenic hemorrhage. *J Trauma*. 2004;57:S29-32

[121] Cole DJ, Connolly RJ, Chan MW, Schwaitzberg SD, Byrne TK, Adams DB, Baron PL, O'Brien PH, Metcalf JS, Demcheva M, Vournakis J. A pilot study evaluating the efficacy of a fully acetylated poly-N-acetyl glucosamine membrane formulation as a topical hemostatic agent. *Surgery*. 1999;126:510-517

[122] Najjar SF, Healey NA, Healey CM, McGarry T, Khan B, Thatte HS, Khuri SF. Evaluation of poly-N-acetyl glucosamine as a hemostatic agent in patients undergoing cardiac catheterization: a double-blind, randomized study. *J Trauma*. 2004;57:S38-41

[123] King DR, Cohn SM, Proctor KG. Modified rapid deployment hemostat bandage terminates bleeding in coagulopathic patients with severe visceral injuries. J *Trauma*. Vol. 57; 2004:756-759

In: Biomaterials Research Advances
Editor: J. B. Kendall, pp. 21-36
ISBN: 978-1-60021-892-7

Chapter 2

MODULATION OF CYCLIC AMP PRODUCTION IN FIBROBLASTS ATTACHED TO SUBSTRATA WITH DIFFERENT SURFACE CHEMISTRIES

E. Bergeron [a], ***E. Lord*** [a], ***M. E. Marquis*** [a], ***T. Groth*** [b] ***and N. Faucheux*** [a,*]

[a] Université de Sherbrooke, Chemical Engineering Department, Sherbrooke, Québec, Canada, J1K 2R1

[b] Institute of Pharmaceutical Engineering, Martin Luther University Halle-Wittenberg, 06099 Halle(Saale), Germany

ABSTRACT

The interactions of cells with biomaterials have been widely studied. However, little is known about the influence of the properties and chemistry of the substratum on the activation of one of the major signaling cascades, the cyclic AMP (cAMP) pathway, in adhering cells. The second messenger cAMP plays a major role in modulating cell morphology, enabling cells to survive, proliferate and differentiate. Our previous study showed that murine Swiss 3T3 fibroblasts loosely attached to a hydrophilic cellulose membrane have a high amount of intracellular cAMP. By contrast, cells well spread on tissue culture polystyrene (PS) contained low concentrations of cAMP. But the surface properties of the cellulose membrane are heterogeneous in terms of rugosity, porosity and chemical composition.

Hence, to study the impact of surface chemistry on Swiss 3T3 fibroblast behaviors, we have prepared self assembled monolayers (SAMs) on glass from alkylsilanes to obtain model surfaces with a variety of terminating functional groups, such as carboxylic acid (COOH), amine (NH_2), poly(ethylene glycol) (PEG) and methyl (CH_3). Our results revealed that the cAMP production was significantly lower in cells attached to COOH-

and NH_2-terminated SAMs than in cells on PEG and CH_3 substrata. Spread cells attached to COOH- and NH_2-terminated SAMs could organize their cytoskeleton, phosphorylate the Tyr^{397} of focal adhesion kinase (FAK) and activate RhoA. By contrast, cells on PEG and CH_3 substrata remained rounded up with few punctuate focal adhesion complexes, while FAK phosphorylation on Tyr^{397} and RhoA activation were partly inhibited. Indeed, an increase in intracellular cAMP severely impaired the formation of focal adhesion complexes and decreased the phosphorylation of Tyr^{397} in FAK. We also found that the phosphorylation of ERK1/2 was also significantly greater in cells attached to COOH and NH_2 substrata than in cells adhering to PEG-terminated SAM. We conclude that monitoring cAMP may contribute to a better understanding of the complex phenomenon of cell-material interactions.

Keywords: signal transduction, alkylsilane, focal adhesion, kinases

INTRODUCTION

Several studies have demonstrated that cell-biomaterial interractions depend on surface properties such as roughness topography, wettability, charge, chemistry and surface energy [1,2]. These characteristics influence the conformation, orientation and quantities of adsorbed adhesion proteins such as vitronectin or fibronectin [3,4]. The adsorbed proteins onto the substratum profoundly affect integrin-receptor binding and subsequent cell adhesive events especially the focal adhesion formation [4,5]. Focal adhesion contacts are flat and elongated structures often located at cell periphery [6,7]. They anchor bundles of actin stress fibers through a plaque made up of many different proteins such as cell membrane integrin receptor, vinculin and phosphotyrosine proteins [8].

Nevertheless, few studies have examined the influence of the substratum surface properties on the activation of early biochemical events, such as the cyclic adenosine 3'-5'monophosphate (cAMP) pathway [9,10]. The second messenger cAMP is involved in a wide range of cell functions, including cell proliferation and motility [11], but the cAMP pathway is especially important in cell adhesion, cytoskeletal structure and focal contact formation [12,13]. An increase in intracellular cAMP induced by β adrenergic agonists or by forskolin, a direct activator of adenylyl cyclase, the enzyme producing cAMP from ATP causes marked morphological changes with a loss of focal adhesion and the fragmentation of actin stress fibers in adherent cells [14,15]. An increase in intracellular cAMP can also severely impair the formation of focal adhesion complexes and decrease the phosphorylation of the tyrosine in focal adhesion kinase (FAK) in several cell types inhibiting the spreading of the cells [11,12,16,17]. FAK plays a key role in mediating integrin signal transduction and becomes autophosphorylated on Tyr^{397} after integrin stimulation [16,17]. Phosphorylated Tyr^{397} represents a binding site for the Src-homology 2-domain of Src family kinases [18,19] The cAMP pathway is also a negative modulator of RhoA synthesis. This small GTPase is involved in the activation of integrins by promoting avidity modulation, a process known as

* Send correspondence to: Nathalie Faucheux, Université de Sherbrooke, Chemical Engineering Department, 2500, Boul. de l'Université, Sherbrooke, Québec, Canada, J1K 2R1; Tel: (1) 819-821-8000, 1343, Fax: (1) 819-821-7955; Email: Nathalie.Faucheux@Usherbrooke.ca

inside-out signaling [20-22]. The phosphorylation of the Ser^{188} of RhoA by cAMP-dependent protein kinase (PKA) is a central event in mediating the cellular effects of cAMP [23]. The translocation of phosphorylated RhoA from the membrane to the cytosol by its binding to guanine nucleotide dissociation inhibitor (GDI) terminates RhoA signaling [23]. Cyclic AMP can also inhibit the signaling of RAS/RAF/MAPK/extracellular signal-regulated kinase (ERK) [24]. ERK phosphorylates many substrates, regulating such cell functions as gene expression, cell morphology, proliferation, differentiation and cell death [24].

Our previous study showed that more cAMP is produced when cells are loosely attached to a hydrophilic cellulose membrane [10]. These cells have a disorganized actin cytoskeleton and subnormal amounts of RhoA in the cell membrane. By contrast, cells spread on tissue culture polystyrene (PS) contain low concentrations of cAMP. Cellulose membrane is a polymeric material which possesses a large degree of surface heterogeneity with regard to the type and distribution of functional groups, surface roughness and porosity. Hence, it is difficult to analyze the influence of the surface composition of these materials on the observed results.

We have therefore used model substrata, self-assembled monolayers (SAMs) of alkyl silanes, to clarify the impact of surface properties like the chemical composition of materials on signal transduction. SAMs have specific terminal functions, such as COOH, NH_2, poly(ethylene glycol) (PEG) and CH_3. Water contact angle measurements have revealed that SAMs terminated with -CH_3 produced hydrophobic surfaces, while those with -NH_2 and -COOH prepared moderately wettable surfaces and those with -PEG created wettable surfaces [25]. We analyzed the effect of these functional groups on the response of Swiss 3T3 fibroblasts by measuring cAMP production, FAK Tyr^{397} phosphorylation and the state of RhoA activation. Total and phosphorylated ERK1/2 MAPK (pERK1/2) were also measured.

Materials and Methods

Preparation of substrata

Glass treatment. Glass microscope coverslips (Superior-Marienfeld, Germany) were cleaned by immersion in freshly prepared piranha solution (3:1 mixture of concentrated H2SO4 and 30% H2O2) for 15 min at room temperature. They were then rinsed exhaustively with distilled water (10 x 6 min) and dried.

Sample preparation. The cleaned glass coverslips were coated with self-assembled monolayers terminating in $COOCH_3$, NH_2, PEG or CH_3, as described previously [25,26]. Briefly, each of these monolayers was prepared by one-step procedures using ethanolic 1% (v/v) 10-carbomethoxydecyldimethylchlorosilane, 1% (v/v) 3-aminopropyldimethylethoxy-silane, 1% (v/v) 2-methoxypolyethyleneoxypropyltrimethoxysilane and 5% (v/v) octadecyldi-methylchlorosilane in hexane (Gelest, Tullytown, USA), overnight at room temperature. The coverslips were rinsed with ethanol, washed with distilled water and dried in air. Carboxylic acid (COOH) groups were generated from $COOCH_3$ by heating the $COOCH_3$-terminated monolayers for 30 min at 100°C in acidified water pH 2.5. They were then rinsed with water, dried for 10 min at 110°C and stored in a vacuum dessicator.

Cell experiments

***Cell culture*.** Swiss 3T3 mouse fibroblast cells (CCL-92™, ATCC, Manassas USA) were grown in Dulbecco's modified Eagle's medium (DMEM, Invitrogen, Burlington, Canada) supplemented with 10% foetal bovine serum (FBS, Sigma, Oakville, Canada) and 1% antibiotic-antimycotic solution (Sigma). Cells were removed by trypsinization (Invitrogen), resuspended in DMEM with 10% FBS and seeded on SAMs at a density of 2 x 10^4 cells/cm^2. Cultures were incubated for 45 min or 120 min at 37 °C in a humidified 5% CO_2 atmosphere.

***cAMP measurements*.** Adherent Swiss 3T3 fibroblasts were incubated with ^{3}H-adenine triphosphate (ATP) (5 μCi/mL) for 2h, removed from the coverslip by trypsinization, suspended in DMEM with 10% FBS and seeded on SAMs and PS at a density of 2 x 10^4 cells/cm^2. The cells were incubated for 45 min, rinsed two times with PBS pH 7.4 and the ^{3}H cAMP in the attached cells was extracted using trichloroacetic acid. The broken cell suspension was centrifuged at 4 000 rpm for 5 min and the intracellular ^{3}H cAMP was separated on a Dowex AG50x8-Alumina column and counted in a scintillation counter. Assays were carried out in triplicate and the results obtained were referred to the number of attached cells determined by counting the non-adherent cells.

***Visualization of the distributions of focal adhesions*.** After incubation for 45 min, the cells attached to SAMs and PS with or without 1μM 8-Bromo-cAMP, an analog of cAMP (Sigma) or 1μM forskolin (Sigma), were fixed by incubation in 3% (w/v) paraformaldehyde in phosphate-buffered saline (PBS) pH 7.4 for 15 min and permeabilized for 5 min with 0.5% (v/v) Triton X-100 in PBS. Non-specific binding sites were blocked by incubating them in PBS containing 1% bovine serum albumin (BSA, Sigma) for 30 min. Cells were immunostained by incubating them with mouse monoclonal antibodies raised against vinculin (Sigma, diluted 1:50). Primary antibody binding was visualized by incubation with a fluorescein (FITC)-conjugated anti-mouse IgG antibodies (Sigma, diluted 1:200). All antibodies were diluted in PBS containing 0.1% BSA and cells were incubated with antibodies for 30 min at room temperature. Filamentous actin (F-Actin) was stained by incubation with BODIPY-phalloidin (Molecular Probes, Oregon, USA, diluted 1:200) for 30 min at room temperature. The coverslips were washed, mounted on glass slides and examined under an epifluorescence microscope (Eclipse TE2000-S, Nikon, Mississauga, Canada) equipped with a 60X oil immersion objective and a Retiga 1300R camera (Nikon).

Western blot analysis of the phosphorylated FAK on Tyr397. Swiss 3T3 cells (2 x 10^4/cm^2) in DMEM with 10% FBS were seeded on COOH-, NH_2-, PEG-, CH_3 -terminated SAMs or PS with or without 8-Bromo-cAMP and incubated for 45 min at 37°C in a 5% CO_2 atmosphere. The cells attached to SAMs were lysed at 4°C in 1 mL of 50 mM Tris-HCl, pH 7.4 containing 10% glycerol (v/v) and protease inhibitors (0.1 mM phenylmethylsulfonylfluoride, 10 μg/mL aprotinin and 10 μg/mL leupeptin, Sigma). The proteins were separated by SDS-PAGE and transferred to nitrocellulose membranes using a Transblot Semi-Dry electrophoretic transfer cell (Hoefer TE70, Amersham Pharmacia, Piscataway, USA). The nitrocellulose membranes were stained with Ponceau red (Sigma) to confirm transfer efficiency and then incubated overnight in a 5% (w/v) solution of non-fat dried milk in PBS Tween 20 0.1% (v/v). The nitrocellulose membranes were washed three times with PBS containing Tween 20 0.1% (v/v) and incubated for 120 min at room temperature with a primary mouse antibody against FAK phosphorylated on Tyr^{397} (Chemicon, Temecula, USA, diluted 1:1 000). The membranes were washed three times with 0.1% (v/v), PBS Tween 20 and bound specific antibody was revealed by incubation with a peroxidase-conjugated anti-mouse second antibody (Sigma, diluted 1:10 000). Immunoreactive bands were visualized by chemiluminescent detection (ECL, Roche Diagnostics, Penzberg, Germany) and exposure to X-Ray film (Kodak, Germany).

Detection of GTP-bound RhoA. Cell extracts (500 μg) were prepared as described above and incubated with glutathione S-transferase fusion protein (GST-Rhotekin) solution (Pierce Biotechnology, Rockford). The N-terminal part of rhotekin binds to GTP-bound RhoA *in vitro*, but not to Rac1 or cdc42 [27]. Activated GTP-bound RhoA was solubilized, resolved by electrophoresis on 12% SDS-polyacrylamide gels and transferred to nitrocellulose membranes (BioRad Laboratories, Mississauga, Canada). The blotted membrane was probed with antisera against RhoA and immunoreactive bands were visualized using a peroxidase-conjugated mouse Ig antibody followed by the ECL reaction and exposure to X-Ray film.

Detection of the phosphorylated ERK1/2. Swiss 3T3 cells in DMEM 10% FBS were seeded on SAMs as described above. The cells attached to SAMs were then lysed at 4°C with a cell lysis kit (Bio-Rad Laboratories). pERK1/2 (Thr^{202}/Tyr^{204}, Thr^{185}/Tyr^{187}) was detected using a BioPlex phospho-ERK1/2 MAPK assay kit (BioRad Laboratories) according to the manufacturer's instruction. Briefly, 50 μL of each cell extract was adjusted to a protein concentration of 200-300 μg/mL and placed in a 96-well filter plate containing bead-conjugated antibody against pERK1/2. The plate was incubated overnight on a platform shaker at 300 rpm at room temperature, washed three time with wash buffer, and the antigen-antibody complexes were visualized with a fluorescently labelled antibody raised against pERK1/2. The ERK1/2 total protein was assayed using the BioPlex assay kit (BioRad Laboratories).

Statistics

All statistical computations were performed with GraphPad Instat®3.00 software (GraphPad Software Inc., San Diego, USA). The Student Newman Keuls multiple comparison test (ANOVA) or a Student *t*-test was applied. Values were considered significantly different if $p < 0.05$.

RESULTS

Production of cAMP in Swiss 3T3 fibroblasts attached to the SAMs

We analyzed the impact of the initial cell-SAM interactions on signal transduction by monitoring the intracellular concentration of cAMP. The cAMP produced after incubation for

45 min was referred to that of cells attached to PS. Swiss 3T3 fibroblasts attached to COOH and NH_2 contained significantly less cAMP than cells on PEG or CH_3 ($p < 0.05$, Fig. 1).

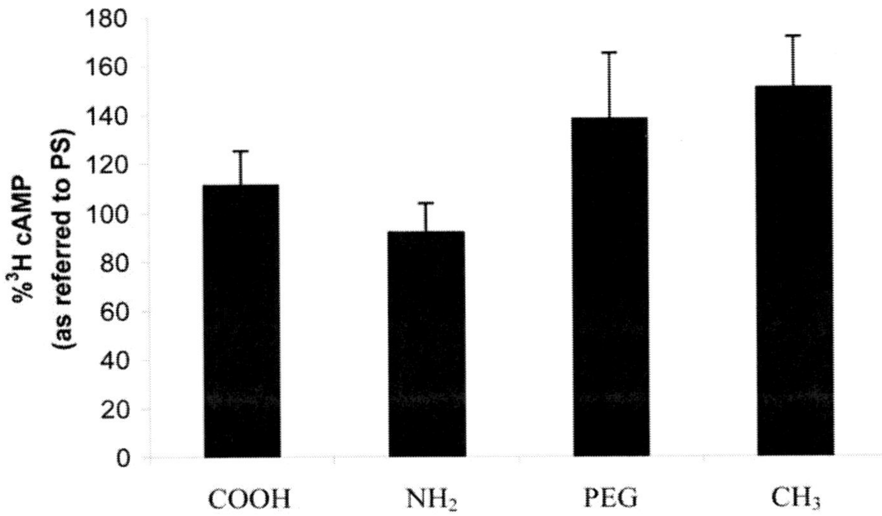

Figure 1: Intracellular cAMP of Swiss 3T3 cells attached to SAMs. The cAMP in cells attached to SAMs was measured as described in *Materials and Methods*. The results are referred to the cAMP in cells attached to PS. Results are the means ± SD of triplicate measurements of a single experiment. Another independent experiment gave similar results.

Organization of the focal adhesions in Swiss 3T3 fibroblasts attached to SAMs

Focal adhesion complexes in adherent cells incubated for 45 min or 120 min (Fig. 2a) were assessed by immunostaining with antibodies against vinculin. F-Actin was also visualized using BODIPY-conjugated phalloidin. Most of the cells attached to COOH substratum had spread little after incubation for 45 min, but they contained some focal adhesion complexes and few stress fibers. On NH_2, the spread cells contained some focal adhesion plaques at the cell periphery and longitudinal actin stress fibers. Swiss 3T3 fibroblasts plated on PEG and CH_3 for 45 min had few focal adhesion complexes and very poorly organized thin stress fibers (Fig. 2a). The cells attached to COOH and NH_2 and incubated for 120 min were fully spread with many bundles of actin stress fibers anchored to the plasma membrane at sites of extended focal adhesion contacts, as demonstrated by intense vinculin clusters (Fig. 2a). By contrast, the cells attached to PEG and CH_3 contained only small focal adhesion complexes at the cell periphery. Most of these cells also remained rounded up and contained few thin stress fibers.

Controls using PS with or without 8-Bromo-cAMP or forskolin (Fig. 2b) revealed that fibroblasts attached to PS after incubation for 45 min at 37°C were well spread. They contained short linear focal adhesion plaques at the cell periphery and longitudinal actin stress fibers. In contrast, few focal adhesions and some actin stress fibers were observed on PS in the presence cAMP analog. Few focal adhesion contacts or actin stress fibers were observed on PS in the presence of forskolin.

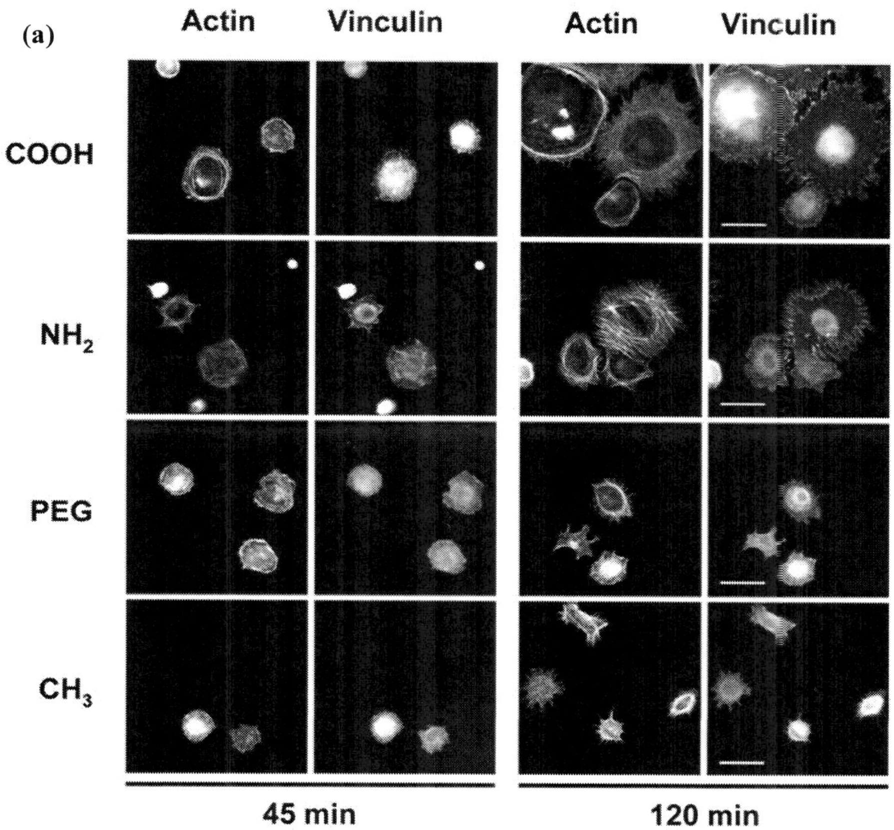

Figure 2: Cytoskeletal organization were visualized by immunostaining of vinculin and labelling of actin in Swiss 3T3 cells attached to SAMs after incubation for 45 min or 120 min **(a)**. Fibroblasts were also seeded on PS in the presence of 1μM 8-Bromo-cAMP or 1μM forskolin (FK) and incubated for 45 min. *(Continue on next page.)*

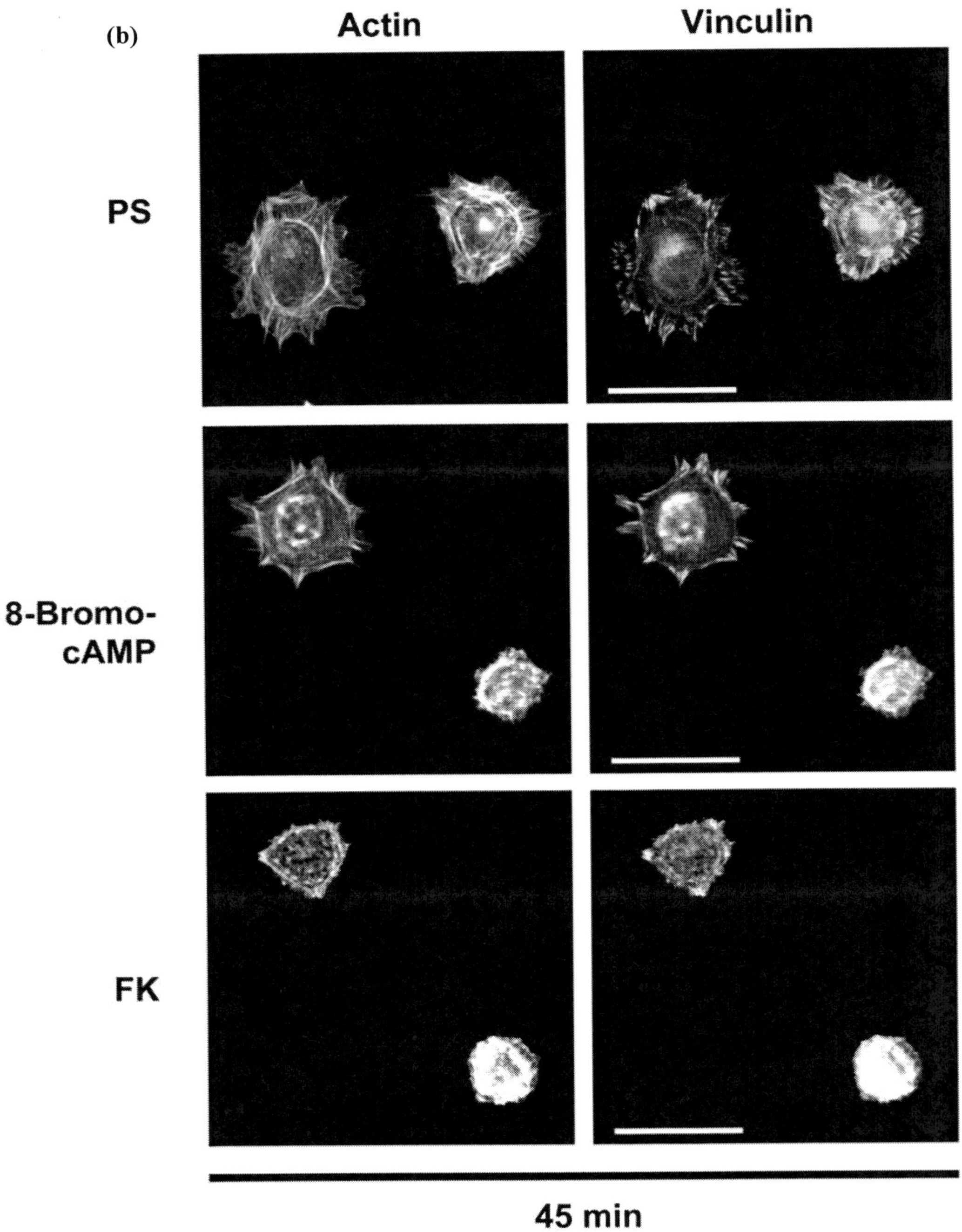

Figure 2: ***Cont.*** **(b)**. The results shown are representative of at least two other experiments. Bar = 50 μm.

Western blot analysis of FAK phosphorylation

The influence of the SAMs on FAK phosphorylation on Tyr^{397} was analyzed by immunoblotting (Fig. 3a) of the proteins in cell extracts prepared from Swiss 3T3 fibroblasts attached to COOH, NH_2, PEG and CH_3 and incubated for 45 min at 37°C. FAK Tyr^{397} phosphorylation was higher in cells on COOH and NH_2 substrata than in cells on PEG and CH_3. The same immunoblot reprobed with antibodies against actin is shown in figure 3a. The various cell extracts contained similar amounts of actin.

Controls using PS with or without 8-Bromo-cAMP (Fig. 3b) revealed that fibroblasts attached to PS after incubation for 45 min contained higher amount of phosphorylated FAK on Tyr^{397} than those attached to PS in the presence of cAMP analog. Control immunoblot probed with antibodies against total FAK is shown in figure 3b. The various cell extracts contained similar amounts of FAK.

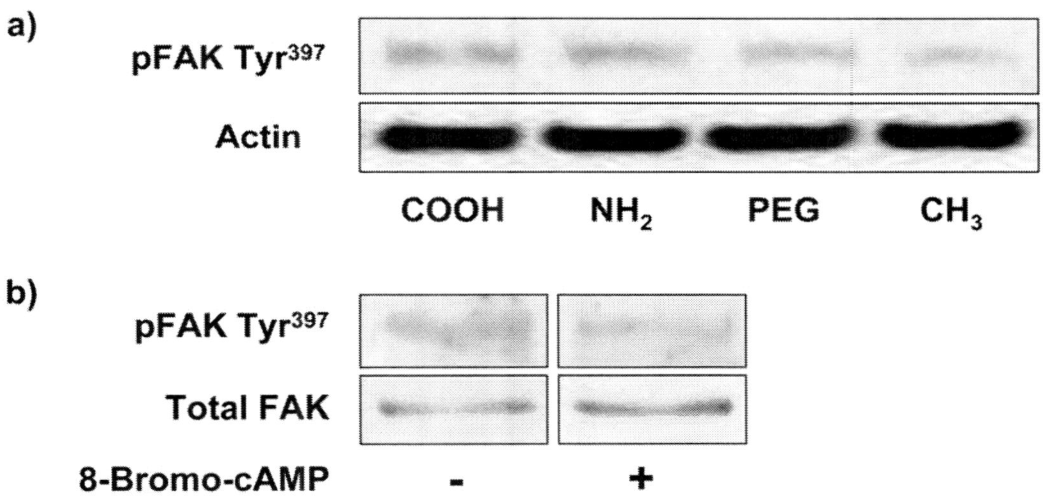

FIgure 3: FAK phosphorylation in Swiss 3T3 cells adhering to COOH-, NH2-, PEG- and CH3-terminated SAMs after incubation for 45 min (a). FAK phosphorylation in fibroblasts seeded on PS in the presence of 1μM 8-Bromo-cAMP (b). Proteins (50 μg) in cell extracts were resolved by SDS PAGE and immunoblotted with monoclonal antibody against FAK phosphorylated on Tyr 397, a monoclonal antibody against actin (a) or a monoclonal antibody against FAK (b), for normalization. The blots shown are representative of two other experiments.

Western blot analysis of GTP-bound RhoA

GTP-bound RhoA was precipitated from extracts (Fig. 4) of Swiss 3T3 fibroblasts attached to COOH, NH_2, PEG and CH_3 and incubated for 45 min at 37°C. The results revealed a major band characteristic of RhoA about 23 kDa [28]. The extracts from Swiss 3T3 cells attached to COOH and NH_2 contained more GTP-bound RhoA than did extracts of cells attached to PEG and CH_3. Control immunoblot probed with antibodies against total RhoA revealed similar amounts of RhoA in all cell extracts (Fig 4).

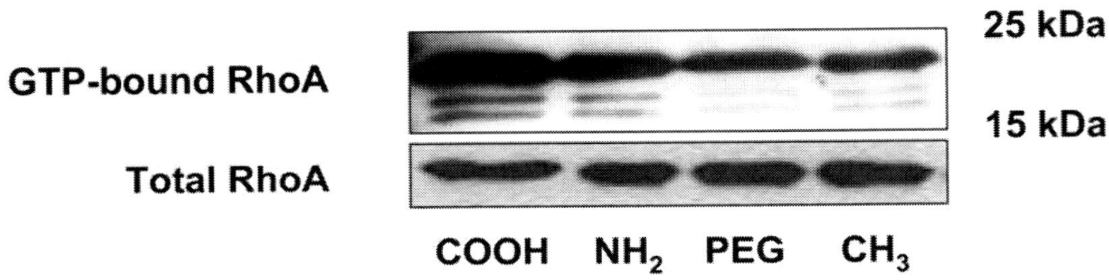

Figure 4: Activation state of RhoA in Swiss 3T3 cells on SAMs. Western blot analysis of GTP-bound RhoA immunoprecipitated by GST-Rhotekin and total RhoA in cell extracts prepared from Swiss 3T3 cells attached to COOH-, NH2-, PEG- and CH3-terminated SAMs. The blots shown are representative of two other experiments showing similar results.

The phosphorylated state of ERK1/2 in cell extracts

pERK1/2 (Thr^{202}/Tyr^{204}, Thr^{185}/Tyr^{187}) was detected in extracts of Swiss 3T3 fibroblasts attached to SAMs after incubation for 45 min at 37°C (Fig 5). As the results for cells on PEG and CH_3 had been similar throughout the study, only the pERK1/2 content of cells on PEG was assessed and not that of cells on CH_3. Each result is expressed relative to the total ERK1/2 content of cells on COOH (11564 ± 2572 fluorescence arbitrary units), NH_2 (13187 ± 1626) or PEG (20342 ± 2387). There was significantly more pERK1/2 in cells on COOH and NH_2 than in cells on PEG ($p < 0.05$).

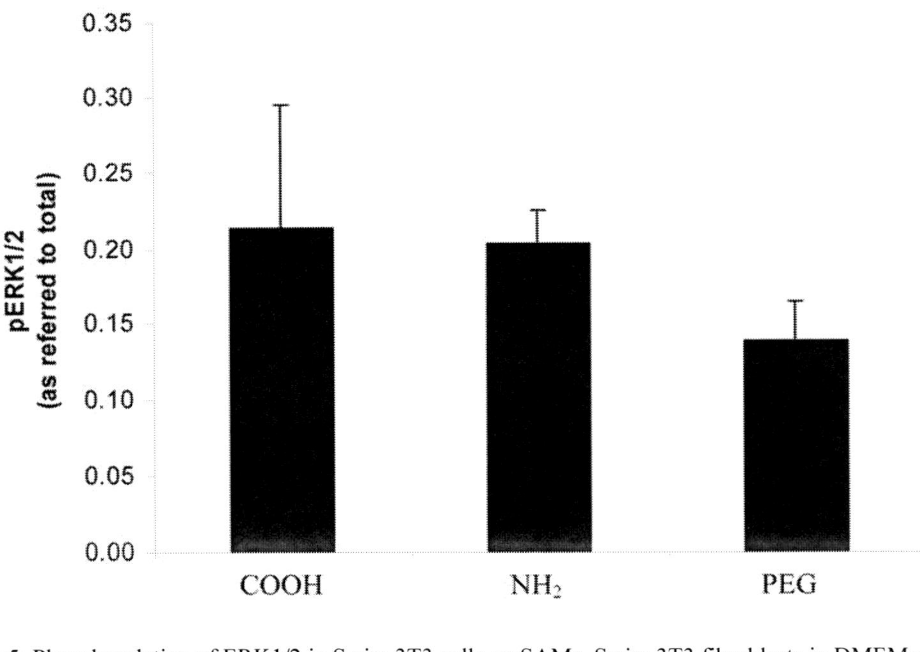

Figure 5: Phosphorylation of ERK1/2 in Swiss 3T3 cells on SAMs. Swiss 3T3 fibroblasts in DMEM plus 10% FBS (2 x 104cells/cm²) were seeded for 45 min at 37°C in 5% CO2 on SAMs. The total and phosphorylated ERK1/2 in cell extracts were quantified as described in Materials and Methods. Results are the means ± SD of six measurements from two independent experiments.

DISCUSSION

The surface properties of substrata control the morphology and behavior of cells that adhere to them [29,30]. Cell adhesion and spreading have been studied on SAMs, well-defined model surfaces with specific terminal groups giving them a low rugosity and controllable wettability [29,31]. Although protein adsorption is the dominant factor regulating cell adhesion [32-34], the way cells perceive information through signal transduction on different substrata is not well understood. The interplay between signals involving cAMP and the cytoskeleton may be especially important in cells attached to biomaterials. The second messenger cAMP can modulate cell morphology regulating cell behaviors [11-15]. We have previously shown that the catalytic activity of adenylyl cyclase, the enzyme that produces cAMP from ATP, is modulated by the attachment of Swiss 3T3 fibroblasts to a cellulose membrane [35]. In this study, we have therefore used SAMs [25,36,37] with a variety of terminating functional groups (COOH, NH_2, PEG and CH_3) to study the cAMP production generated by cell-substrata interactions.

Cells attached to moderately wettable surfaces (COOH and NH_2) and incubated for 45 min can spread and organize their cytoskeleton and contain less cAMP than round cells on highly hydrophilic (PEG) or hydrophobic (CH_3) substrata. Several studies have shown that moderately wettable surfaces support the formation of focal adhesions, while hydrophobic and hydrophilic surfaces inhibit the formation of these structures [38-40]. Although COOH, NH_2 and CH_3 substrata all adsorb vitronectin from FBS [25], only COOH- and NH_2-terminated SAMs support the formation of focal adhesions. Indeed, the conformations of adsorbed proteins onto CH_3 substratum are changed, and this inhibits the adhesion of cells via integrins [41,42]. In contrast, PEG-terminated SAM does not adsorb adhesive proteins [25]. We have recently shown that a lack of binding between adhesive proteins and the integrins activates the cAMP pathway [9]. Here, we showed that cells on PEG- and CH_3-terminated SAMs contained a higher amount of cAMP and lacked well-developed focal adhesions and actin polymerization, indicating impaired adhesive protein-integrin binding, even after incubation for 120 min. An increase in intracellular cAMP can also prevent the formation of focal adhesion [13,14]. Indeed, we showed in this study that Swiss 3T3 fibroblasts attached to PS in the presence of cAMP analog or activator of the adenylyl cyclase contained only few focal adhesion plaques and some actin stress fibers.

We have also analyzed the phosphorylation state of FAK Tyr^{397}. FAK plays a major role in integrin-mediated signal transduction and is autophosphorylated on Tyr^{397} when cells bind to material surfaces [43-45]. Our results reveal that fibroblasts attached to COOH and NH_2 substrata contain similar quantities of phosphorylated FAK on Tyr^{397}, while the phosphorylation of this site is partly inhibited in cells on PEG- and CH_3-terminated SAMs. Using MC3T3-E1 preosteoblast-like cells on fibronectin-coated SAMs, Keselowski *et al.* [46] showed that FAK Tyr^{397} is most highly phosphorylated in cells on fibronectin-coated NH_2 substrata, closely followed by COOH substratum. In contrast, the FAK Tyr^{397} of cells on fibronectin-coated CH_3-terminated SAM is minimally phosphorylated [46]. Here, we showed that the presence of an analog of cAMP drastically reduced the phosphorylation of FAK on Tyr^{397} in Swiss 3T3 attached to PS after incubation for 45 min. Padmanabhan *et al.* [16] have also demonstrated that treating cells with the cAMP analog dibutyryl cAMP decreases FAK phosphophorylation in astrocytes. Thus, cAMP production in cells attached to PEG and CH_3-terminated SAM might be involved in the decreased FAK phosphorylation.

The Rho family of small GTPases, which includes Rho, Rac, and cdc42, plays also a role in the formation of focal adhesions and their maintenance [47]. We have directly measured the amount of GTP-bound RhoA, the activated form of RhoA, in Swiss 3T3 cells attached to the COOH- and NH_2-terminated SAMs. These cells contained more GTP-bound RhoA than do cells on PEG- and CH_3-terminated SAMs. McClary *et al.* [48] have shown that the membranes of spread cells attached to COOH-terminated SAM contained a higher concentration of RhoA than do cells on CH_3-terminated SAM. The cAMP-dependent PKA mediates the phosphorylation of RhoA in cytotoxic T lymphocytes, leading to an inhibition of RhoA [23]. Furthermore, increased cAMP production decreases the amount of RhoA in the membranes of Swiss 3T3 fibroblasts on a hydrophilic cellulosic membrane [10]. This effect is mediated through the PKA since an inhibitor of the kinase (PKI) can restore RhoA at the cell membrane [10]. In addition, Swiss 3T3 fibroblasts attached to PS in the presence of forskolin or an analog of cAMP contained a low amount of GTP-RhoA in comparaison to PS without treatment (data not shown). Therefore, the cAMP production in Swiss 3T3 fibroblasts attached to both hydrophilic PEG and hydrophobic CH_3 substrata may mediate the inhibition of RhoA.

GTP-bound RhoA, which plays a critical role in the spreading of the Swiss 3T3 fibroblasts, is also involved in the activation of the ERK1/2 MAPK pathway [49]. This pathway is activated by the binding of growth factors to the tyrosine kinase receptors. This leads to the activation of the small G-protein RAS. Then, c-RAF, MEK and ERK1/2 are activated in a cascade of phosphorylation events [49].

Cyclic AMP inhibits the ERK pathway in several cell types, including adipocytes, fibroblasts [49,50] and smooth muscle cells [51]. The cell-permeable cAMP analog 8-Chloro-cAMP also blocks the ERK pathway in fibroblasts [21,49,52], showing that there is crosstalk between these pathways. Our results indicate that ERK1/2 is more highly phosphorylated in cells on COOH and NH_2 substrata than in cells on a PEG-terminated SAM. It is generally agreed that the inactivation of ERK1/2 by the cAMP pathway is due to inhibition of c-RAF, but the precise mechanism is not clear. This crosstalk may imply that there are other intracellular targets, since cAMP still inhibits ERK, even when c-RAF signaling is restored [49,52].

Conclusion

Few studies have investigated the influence of surfaces bearing different terminating groups on the early biochemical events generated by cell-substratum interactions. A better knowledge of these phenomena is required for the development of new biomaterials that produce specific cell responses. This study of the early steps of cell-material adhesions (Table 1) highlights variations in cAMP produced by Swiss 3T3 fibroblasts on SAMs with different surface groups and may be associated with states of FAK, RhoA and ERK1/2 activation. Thus, monitoring the early intracellular concentration of cAMP may indicate the quality of the cell-substratum interactions controlling the subsequent signal transduction and therefore cell behavior.

Table 1: Cytoskeletal organization and signal transduction of Swiss 3T3 cells on SAMs.

Surface Properties	*Functional End Groups*	*cAMP Production*	*Cytoskeleton Organization*	*FAK Phosphorylation on Tyr^{397}*	*GTP-Bound RhoA*	*ERK 1/2 Phosphorylation*
Moderate Wettable	COOH	+/-	++	+++	+++	++
Moderate Wettable	NH_2	+/-	++	+++	++	++
Hydrophilic	PEG	++	+/-	++	+	+
Hydrophobic	CH_3	++	+/-	+	+	NA

ACKNOWLEDGEMENT

We thank C. CORREZE (INSERM U486, Endocrinologie, Faculté de Pharmacie de Châtenay-Malabry, France) for help in measuring cAMP production. The English text was edited by Dr Owen Parkes.

REFERENCES

[1] Mahmood, T.A., de Jong, R., Riesle, J., Langer, R. and van Blitterswijk, C.A. Adhesion-mediated signal transduction in human articular chondrocytes: the influence of biomaterial chemistry and tenascin-C. *Exp. Cell Res.* 301, 179, 2004.

[2] Diener, A., Nebe, B., Luthen, F., Becker, P., Beck, U., Neumann, H.G. and Rychly, J. Control of focal adhesion dynamics by material surface characteristics. *Biomaterials*. 26, 383, 2005.

[3] Wilson, C.J., Clegg, R.E., Leavesley, D.I. and Pearcy, M.J. Mediation of biomaterial-cell interactions by adsorbed proteins: a review. *Tissue Eng.* 11, 1, 2005.

[4] Garcia, A.J. Get a grip: integrins in cell-biomaterial interactions. *Biomaterials*. 26, 7525, 2005.

[5] Keselowsky, B.G., Collard, D.M. and Garcia, A.J. Integrin binding specificity regulates biomaterial surface chemistry effects on cell differentiation. *Proc. Natl. Acad. Sci.* U.S.A. 102, 5953, 2005.

[6] Sastry, S.K. and Burridge, K. Focal adhesions : a nexus for intracellular signaling and cytoskeletal dynamics. *Exp. Cell Res.* 261, 25, 2000.

[7] Volberg, T., Romer, L., Zamir, E. and Geiger, B. pp60src and related tyrosine kinases: a role in the assembly and reorganization of matrix adhesions. *J. Cell Science.* 114, 2279, 2001.

[8] Geiger, B., Bershadsky, A., Pankov, R. and Yamada, K.M. Transmembrane extracellular matrix-cytoskeleton crosstalk. *Nat Rev Mol Cell Biol.* 2, 793, 2001.

[9] Faucheux, N., Haye, B. and Nagel, M.D. Activation of the cyclic AMP pathway in cells adhering to biomaterials: regulation by vitronectin- and fibronectin-integrin binding. *Biomaterials*. 21, 1031, 2000.

[10] Faucheux, N. and Nagel, M.D. Cyclic AMP-dependent aggregation of Swiss 3T3 cells on a cellulose substratum (Cuprophan) and decreased cell membrane RhoA. *Biomaterials*. 23, 2295, 2002.

[11] Lampugnani, M.G., Giorgi, M., Gaboli, M., Dejana, E. and Marchisio, P.C. Endothelial cell motility, integrin receptor clustering and microfilament organization are inhibited by agents that increase intracellular cAMP. *Lab. Invest*. 63, 521, 1990.

[12] Lamb, N.J.C., Fernandez, A., Conti, M.A. Adelstein, R., Glass, D.B., Welch, W.J. and Feramisco, J.R. Regulation of actin microfilament integrity in living nonmuscle cells by the cAMP-dependent protein kinase and the myosin light chain kinase. *J. Cell Biol.* 106, 1955, 1988.

[13] Han, J.D. and Rubin, C.S. Regulation of cytoskeleton organization and paxillin dephosphorylation by cAMP. *J. Biol. Chem*. 271, 29211, 1996.

[14] Willingham, M.C. and Pastan, I. Cyclic AMP and cell morphology in cultured fibroblasts. *J. Cell Biol*. 67, 146, 1975.

[15] Petty, H.R. and Martin, S.M. Combinative ligand-receptor interactions : effects of cAMP, epinephrin and met enkephalin on RAW 264 macrophage morphology, spreading, adherence and microfilaments. *J. Cell. Physiol*. 138, 247, 1989.

[16] Padmanabhan, J., Clayton, D. and Shelanski, M.L. Dibutyryl cyclic AMP-induced process formation in astrocytes is associated with a decrease in tyrosine phosphorylation of focal adhesion kinase and paxillin. *J. Neurobiol.* 39, 407, 1999.

[17] Troyer, D.A., Bouton, A., Bedolla, R. and Padilla, R. Tyrosine phosphorylation of focal adhesion kinase (p125(FAK)): Regulation by cAMP and thrombin in mesangial cells. *J. Am. Soc. Nephrol.* 7, 415, 1996.

[18] Hildebrand, J.D., Schaller, M.D. and Parsons, J.T. Identification of sequences required for the efficient localization of the focal adhesion kinase, pp125FAK, to cellular focal adhesions. *J. Cell Biol.* 123, 993, 1993.

[19] Chen, H.C., Appeddu, P.A., Parsons, J.T., Hildebrand, J.D., Schaller, M.D. and Guan, J.L. Interaction of focal adhesion kinase with cytoskeletal protein talin. *J. Biol. Chem*. 270, 16995, 1995.

[20] Barry, S.T., Flinn, H.M., Humphries, M.J., Critchley, D.R. and Ridley, A.J. Requirement for Rho in integrin signalling. *Cell Adhes. Commun.* 4, 387, 1996.

[21] Burgering, B.M., Pronk, G.J., van Weeren, P.C., Chardin, P. and Bos, J.L. cAMP antagonizes p21Ras-directed activation of extracellular signal-regulated kinase 2 and phosphorylation of mSos nucleotide exchange factor. *EMBO J*. 12, 4211, 1993.

[22] Ridley, A.J. and Hall, A. The Small GTP-Binding Protein Rho Regulates the Assembly of Focal Adhesions and Actin Stress Fibers in Response to Growth-Factors. *Cell*. 70, 389, 1992.

[23] Lang, P., Gesbert, F., Delespine-Carmagnat, M., Stancou, R., Pouchelet, M. and Bertoglio, J. Protein kinase A phosphorylation of RhoA mediates the morphological and functional effects of cyclic AMP in cytotoxic lymphocytes. *EMBO J.* 15, 510, 1996.

[24] Dumaz, N. and Marais, R. Integrating signals between cAMP and the RAS/RAF/MEK/ERK signalling pathways. *FEBS J*. 272, 3491, 2005.

[25] Faucheux, N., Schweiss, R., Lutzow, K., Werner, C. and Groth, T. Self-assembled monolayers with different terminating groups as model substrates for cell adhesion studies. *Biomaterials*. 25, 2721, 2004.

[26] Faucheux, N., Tzoneva, R., Nagel, M.D. and Groth, T. The dependence of fibrillar adhesions in human fibroblasts on substratum chemistry. *Biomaterials*. 27, 234, 2006.
[27] Reid, T., Furayashiki, T., Ishizaki, T., Watanabe, G., Watanabe, N., Fujisawa, K., Morii, N., Madaule, P. and Narumiya, S. Rhotekin, a new putative target for Rho bearing homology to a serine/threonine kinase, PKN, and rhophilin in the Rho-binding domain. *J. Biol. Chem.* 271, 13556, 1996.
[28] Bilodeau, D., Lamy, S., Desrosiers, R.R., Gingras, D. and Beliveau, R. Regulation of Rho protein binding to membranes by rhoGDI: inhibition of releasing activity by physiological ionic conditions. *Biochem. Cell. Biol.* 77, 59, 1999.
[29] Healy, K.E., Lom, B. and Hockberger, P.E. Spatial Distribution of Mammalian Cells Dictated by Material Surface Chemistry. *Biotech. Bioeng.* 43, 792, 1994.
[30] Ito, Y. Surface micropatterning to regulate cell functions. *Biomaterials*. 20, 2333, 1999.
[31] Sukenik, C.N., Balachander, N., Culp, L.A., Lewandowska, K. and Merritt, K. Modulation of cell adhesion by modification of titanium surfaces with covalently attached self-assembled monolayers. *J. Biomed. Mater. Res.* 24, 1307, 1990.
[32] Balcells, M. and Edelman, E.R. Effect of pre-adsorbed proteins on attachment, proliferation, and function of endothelial cells. *J. Cell. Physiol.* 191, 155, 2002.
[33] Goldstein, A.S. and DiMilla, P.A. Effect of adsorbed fibronectin concentration on cell adhesion and deformation under shear on hydrophobic surfaces. *J. Biomed. Mater. Res.* 59, 665, 2002.
[34] Curtis, A.S. and Forrester, J.V. The competitive effects of serum proteins on cell adhesion. *J. Cell. Sci.* 71, 17, 1984.
[35] Faucheux, N., Correze, C., Haye, B. and Nagel, M.D. Accumulation of cyclic AMP in Swiss 3T3 cells adhering to a cellulose biomaterial substratum through interaction with adenylyl cyclase. *Biomaterials*. 22, 2993, 2001.
[36] Mrksich, M. and Whitesides, G.M. Using self-assembled monolayers to understand the interactions of man-made surfaces with proteins and cells. *Annu. Rev. Biophys. Biomol. Struct.* 25, 55, 1996.
[37] Schreiber, F. Structure and Growth of Self-Assembling Monolayers. *Prog. Surf. Sci.* 65, 151, 2000.
[38] McFarland, C.D., Thomas, C.H., DeFilippis, C., Steele, J.G. and Healy, K.E. Protein adsorption and cell attachment to patterned surfaces. *J. Biomed. Mater. Res.* 49, 200, 2000.
[39] Groth, T. and Altankov, G. Insights into the tissue compatibility of biomaterials. In: Proc. 9th International Conference on Polymers in Medecine and Surgery. IOM Communications, London, UK: Chameleon Press Limited, 2000, pp. 205-213.
[40] McClary, K.B., Ugarova, T. and Grainger, D.W. Modulating fibroblast adhesion, spreading, and proliferation using self-assembled monolayer films of alkylthiolates on gold. *J. Biomed. Mater. Res.* 50, 428, 2000.
[41] Banovac, F., Saavedra, S.S. and Truskey, G.A. Local Conformational Changes to Vitronectin Upon Adsorption to Glass and Silane Surfaces. *J. Colloid Interface Sci.* 165, 31, 1994.
[42] Iuliano, D.J., Saavedra, S.S. and Truskey, G.A. Effect of the conformation and orientation of adsorbed fibronectin on endothelial cell spreading and the strength of adhesion. *J. Biomed. Mater. Res.* 27, 1103, 1993.

[43] Guan, J.L. and Shalloway, D. Regulation of focal adhesion-associated protein tyrosine kinase by both cellular adhesion and oncogenic transformation. *Nature*. 358, 690, 1992.

[44] Schaller, M.D., Hildebrand, J.D., Shannon, J.D., Fox, J.W., Vines, R.R. and Parsons, J.T. Autophosphorylation of the focal adhesion kinase, pp125FAK, directs SH2-dependent binding of pp60src. *Mol. Cell. Biol.* 14, 1680, 1994.

[45] Hanks, S.K., Calalb, M.B., Harper, M.C. and Patel, S.K. Focal adhesion protein-tyrosine kinase phosphorylated in response to cell attachment to fibronectin. *Proc. Natl. Acad. Sci. U.S.A.* 89, 8487, 1992.

[46] Keselowsky, B.G., Collard, D.M. and Garcia, A.J. Surface chemistry modulates focal adhesion composition and signaling through changes in integrin binding. *Biomaterials*. 25, 5947, 2004.

[47] Hall, A. Rho GTPases and the actin cytoskeleton. *Science*. 279, 509, 1998.

[48] McClary, K.B. and Grainger, D.W. RhoA-induced changes in fibroblasts cultured on organic monolayers. *Biomaterials*. 20, 2435, 1999.

[49] Hotchin, N.A. and Hall, A. The assembly of integrin adhesion complexes requires both extracellular matrix and intracellular rho/rac GTPases. *J. Cell Biol.* 131, 1857, 1995.

[50] Sevetson, B.R., Kong, X. and Lawrence Jr, J.C. Increasing cAMP attenuates activation of mitogen-activated protein kinase. *Proc. Natl. Acad. Sci. U.S.A.* 90, 10305, 1993.

[51] Osinski, M.T. and Schror, K. Inhibition of platelet-derived growth factor-induced mitogenesis by phosphodiesterase 3 inhibitors: role of protein kinase A in vascular smooth muscle cell mitogenesis. *Biochem. Pharmacol.* 60, 381, 2000.

[52] Cook, S.J. and McCormick, F. Inhibition by cAMP of Ras-dependent activation of Raf. *Science*. 262, 1069, 1993.

In: Biomaterials Research Advances
Editor: J. B. Kendall, pp. 37-66
ISBN: 978-1-60021-892-7

Chapter 3

THE BEHAVIOR OF ENDOTHELIAL CELLS IN 3D BIOMATERIALS FOR TISSUE ENGINEERING APPLICATIONS

Amit Jairaman and Shan-hui Hsu*
Department of Chemical Engineering, National Chung-Hsing University, Taichung- 402, Taiwan R.O.C

ABSTRACT

Endothelial cells (EC) play a vital role in tissue engineering (TE) - ranging from the design of small- to medium-sized tissue engineered blood vessel (TEBV) constructs to the creation of micro-vascular networks essential for the supply of oxygen and nutrients to the three-dimensional (3D) tissue assemblies. The first part of the study compared the behavior of bovine aortic arterial endothelial cells (BEC) cultured in a 3D gelatin scaffold having two different pore sizes, with that of the conventional 2D culture. DNA assay, PI staining, SEM and RT-PCR were done to evaluate the behavior of EC in 3D culture conditions. Specific emphasis was laid on the effect of pore size on EC behavior. The second part of the study evaluated BEC following treatment with low energy laser irradiation (LELI) from a diode laser. Recent work has focused on enhancing EC functions by the physical stimulation such as cyclic mechanical stress. The relatively few studies on the effect of low energy laser irradiation (LELI) on EC have been mostly been done on venous EC and have used He-Ne (helium neon) lasers. So BEC cultures were treated with LELI- having different energies and for different time periods. MTT tests, propidium iodide (PI) staining followed by FACS analysis and RT-PCR tests were done to determine cell viability, cell-cycle profiles and endothelial nitric oxide synthase (eNOS) gene expression. An increase in the proliferation rates and gene expression was observed at certain specific intensities. The comparative study of EC in 3D and 2D cultures in this study may provide some valuable background information that might useful in the future evaluation of EC for TE applications. An increase in the eNOS gene expression following LELI may be of potential benefit in the use of EC for various applications especially for EC in 3D biomaterials

A. Endothelial Cell Behavior on a Novel 3D Gelatin Scaffold Crosslinked with Genipin.

1. Introduction

End-stage organ failure or tissue loss is one of the most devastating and expensive problems in medicine. Over the last 50 years, transplantation of a wide variety of tissues, reconstructive surgical techniques, and replacement with mechanical devices have significantly improved patient outcomes [1,2]. While great strides have been made, organ and tissue transplantation are imperfect solutions because they are limited by a number of factorslike lack of sufficient availability of donor tissue, life long dependence on immunosupression with its impending complications etc [1]. Because of the above shortcomings, the field of tissue engineering and selective cell transplantation was born as a means to replace diseased tissue with living tissue that is "designed and constructed to meet the needs of each individual patient" [3]. Tissue engineering is "an interdisciplinary field that applies the principles and methods of engineering and the life sciences toward the development of biological substitutes that restore, maintain, or improve tissue function" [1,3].The goal of tissue engineering is to "restore function through the delivery of living elements which become integrated into the patient" [3].

One of the major limitations to the success of tissue engineering has been the difficulties encountered in the development of a viable blood supply to the three dimensional (3D) tissue constructs [4].This involves the creation of both a micro vascular network and a macroscopic circulation and is critical because cells can stay alive by diffusion only when they are within 150–200 uM of a blood supply [5,6].Overcoming the above challenge has been one of the prime goals in this field.

Both angiogenesis and vasculogenesis, in a series of ordered molecular events, play a vital role in the development of such vascular networks [7-9]. The endothelial cells, being the major components of the vascular system coordinate vascular development by responding to various factors such as perfusion rate, shear stress, and low oxygenation, as well as to many angiogenic growth signals (VEGF, b-FGF etc) to finally produce a mature vascular network [10].

The development of independently vascularised artificial tissue also requires a matrix through which exuding nutrients can perfuse cells [11]. The matrix can also function as a reservoir of growth factors to induce incoming blood vessel (angioinduction) as well as a scaffold to seed cells like endothelial cells (ECs) and endothelial progenitor cells (EPCs), which participate in the formation of newer blood vessels by the process of arteriogenesis. Micro vessel networks, artificial matrices and neovascularisation thus constitute the triad of tissue vascularisation .

Relatively few studies have addressed endothelial cell tube formation on synthetic biomaterials. Among the materials used for such studies, collagen, matrigel, fibrin, PLGA etc has been studied more extensively [11]. Matsuda et al showed that after 2 to 3 weeks of culture, endothelial cells cultured on the intermediately hydrophobic cellulosic surfaces, but not other surfaces, formed tube-like structures [12]. Generally, endothelial cells cultured on basement membrane ECM components (collagen IV and laminin) formed tubular structures sooner (hours to days) than those cultured on other natural or synthetic materials, on which cells took days to weeks to form tubes [13]. VEGF incorporated into polyglycolic acid-poly-L-lactic acid (PGA-PLLA) matrices was found to increase the in growth of micro vessels from the host vasculature [14]. Xiao et al showed that EPC-derived EC seeded with human smooth muscle cells (SMC) formed capillary-like structures throughout the PLLA-PGA scaffold [15]. However, seeding EC alone into the scaffold without SMC did not form any such micro vessels even over a large time period. In spite of many such works being done to improve micro vessel formation, there have been relatively few *in vitro* studies on basic biomaterial- endothelial interactions in a 3D environment.

Therefore the basic aim of this project was to make a very basic study on the behavior of endothelial cells on a 3D scaffold (without the incorporation of growth factors). As a first step, this included selecting a suitable 3D scaffold favorable to EC growth followed by the characterization of such a scaffold by microscopy, mechanical and degradation tests. While collagen forms a major component of the endothelial basement membrane and makes for a good 3D scaffold, it is of animal origin and is antigenic. Gelatin on the other hand is denatured collagen and has relatively low antigenicity and has good biocompatibility and biodegradability [16,17]. Moreover, gelatin is conventionally used as a coating in tissue culture flasks to culture ECs as it provides an ideal environment to maintain the function of the ECs. Therefore, it was decided to use gelatin to make 3D scaffolds.

However, the main limitation of gelatin for the preparation of tissue substitutes is its rapid dissolution in aqueous environments, leading to fast degradation of grafts at body temperature [18]. While chemical cross linking with agents like formaldehyde solves the problem, such agents can be unfavorable to the growth of cells. Genipin, a plant based product widely used in herbal medicine, can form stable crosslinked products resistant to enzymatic degradation [19,20] and moreover, gelatin-bioadhesives crosslinked with genipin display higher biocompatibility and less cytotoxicity than other agents [21]. So it was decided to use genipin to crosslink the scaffold to enhance its mechanical strength.

The final part of the project included seeding of ECs into these 3D scaffolds and observing their morphology and function as compared to a more conventional monolayer environment. Thus, the basic aim of the work was to evaluate EC behavior in 3D andcompare such behavior with conventional 2D behavior thus providing some valuable background information that might useful in the future evaluation of ECs for tissue engineering applications.

2. Materials and Methods

2.1 Synthesis of pure gelatin scaffolds

Type A gelatin, 300 bloom extracted from porcine skin (mol. wt 50-100 x 10^3) was purchased from *Sigma chemical co*. The synthesis of gelatin scaffolds was done by a freeze-dry method [22,23]. Three different concentrations of gelatin solution were prepared in distilled water (2.5%, 3.5% and 5 %) and stored at 60 deg. C for 1 hour. The solutions were poured into non-tissue culture petri dishes, purchased from *keonig*, to obtain scaffolds of the required diameter (5.5 cm) and thickness (3 mm). For each concentration of gelatin used, a few samples were stored at - 20 deg. C and a few at - 80 deg. C so as to obtain scaffolds of two different pore sizes. Following this, all samples were immediately transferred to a freeze dry equipment maintaining a temperature of -50 deg. C and a pressure of 0 torr (vacuum), for 24 hours to obtain the final scaffold.

2.2 Cross linking of gelatin scaffolds

98% pure genipin (mol. wt 226.23) was obtained from *Sigma*. Two different concentrations of genipin solution (0.5 and 1%) were prepared in distilled water. The solution was then poured into petri dishes containing the freeze-dried gelatin scaffolds and the gelatin-genipin cross-linking reaction was allowed to proceed at 25 deg. C for 12 to 24 hours with gentle shaking. Following the reaction, scaffolds were washed initially with 22% alcohol to remove the unreacted genipin followed by distilled water for 26 hours to remove the residual alcohol. The crosslinked samples were kept at -20 deg. C for 12 hours and then freeze dried again for 24 hours to obtain the final cross- linked scaffold [22,23].

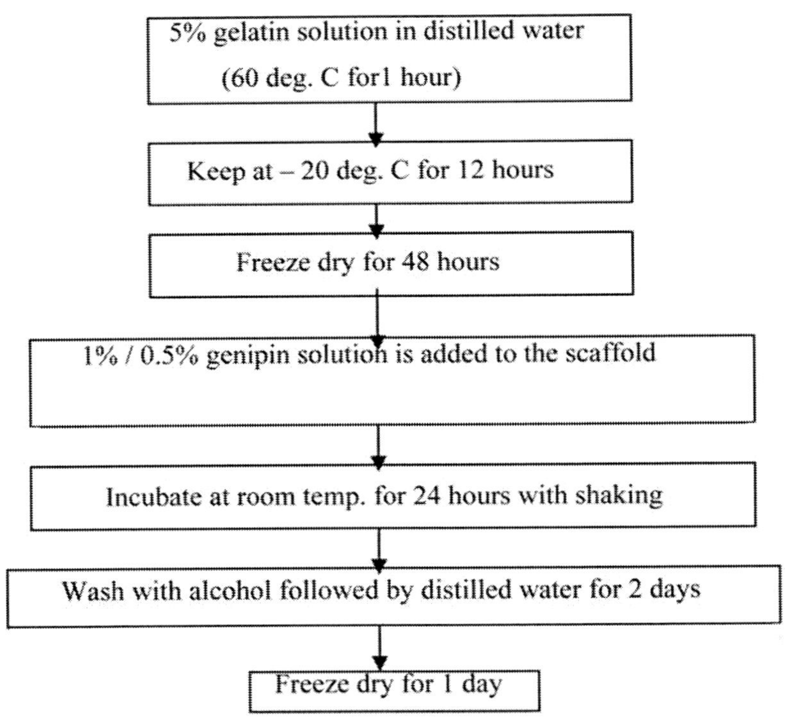

Figure 1: Preparation of gelatin scaffolds.

2.3 Scaffold morphology

The scaffold morphology was observed by optical microscopy and electron microscopy . The following different scaffold types were observed: 2.5%, 3.5% and 5% gelatin- each freeze dried at -20 deg. C and -80 deg. C respectively. The morphology was observed with regard to pore size and pore distribution. After selecting an appropriate gelatin scaffold for experiments, the scaffold morphology was observed under the optical microscope following cross linking with 0.5% and 1% genipin respectively- in both dry and wet conditions (in distilled water and in medium).

2.4 Swelling ratio

The swelling ratio of a scaffold determines its capacity to swell following absorption of water and is an important parameter for scaffold use. The swelling ratio of gelatin scaffolds was investigated as a function of the genipin concentration (0.5% and 1%) and the cross-linking time. This was compared with that of pure gelatin used as control. The dry weight of each scaffold was determined. Following this, the scaffolds were placed in DMEM medium. The wet weights of the scaffolds were then determined at different time points. The swelling ratio was calculated using the equation:

$$E\ (\%) = ((Ws - W_d) / W_d) \times 100\ [24]$$

where E is the water absorption (%wt) of the films, W_d and Ws are the weights of the samples in the dry and swollen states respectively.

2.5 Degradation test

The enzyme bacterial collagenase type II was used to test the degradation of pure and cross linked gelatin scaffolds [25]. Collagenase enzyme was used at a concentration of 31.2 U/ml (0.1 mg/ml PBS). The dry weight was determined for each scaffold, and following treatment with collagenase at different time points, the scaffolds were freeze dried for 24 hours and the dry weights determined again. The weight loss percentage was used as an indicator of the scaffold degradation rate.

2.6 Endothelial cell culture

Bovine aortic arterial endothelial cells (BEC) having passage numbers 15 to 25 were used for the experiments. The cells were cultured in H-DMEM (High glucose Dulbecco's Modified Eagle medium) from *Gibco Laboratories* containing L-Glutamine, pyridoxal hydrochloride, Sodium pyruvate, Sodium bicarbonate and supplemented with 10% FBS (fetal bovine serum) from *Biological Industries*. Penicillin- streptomycin was used as antibiotic. BEC was incubated at 37 deg. C in a standard air/CO_2 incubator containing 5% CO_2. For subculture, the BEC were washed twice with PBS (phosphate buffered saline), trypsinised using 0.05% trypsin-EDTA from *Gibco laboratories* and seeded onto gelatin scaffolds in 24-well plates for experiments.

2.7 Seeding ECs into the 3D scaffold

Seeding of ECs into the 3D scaffolds was done by two methods.

Method 1: The scaffolds were cut into small sizes having 6mm diameter and 2.5 mm thickness each. They were placed in 24 well plates. The cell suspension containing about 3 x 10^5 cells/ 100 μL was added on to the scaffolds. The 24 well plates were then kept in a CO_2 incubator at 37 deg. C.

Method 2: A suspension of BEC containing 3x 10^6 cells was taken in a 15 ml falcon tube. The scaffolds were cut into small sizes having 6mm diameter and 2.5 mm thickness each and 3 such scaffolds were added to each tube. The tubes were shaken at a high speed in a shaker kept inside a 37 deg. C incubator for 2 hours to facilitate cell penetration and attachment into the scaffold. The scaffold was then transferred to 24 well plates and H-DMEM medium was added onto the scaffolds.

2.8 DNA assay for cell quantification

The cell number within the scaffolds was determined by using a fluorescence based method called DNA assay. Briefly, the scaffolds seeded with cells were lyophilized, degraded with a solution containing papain, and then treated with Hoechst dye that binds to DNA. The amount of dye absorbed by each sample was determined by a fluorescent spectrometer, thus indicating the cell number based on a standard curve [26].

2.9 Cell morphology using SEM

To study cell morphology within the 3D gelatin scaffolds, the scaffold containing cells were fixed in 2.5% glutaraldehyde solution and stored at 4 deg. C for 24 hours. The samples were then serially dehydrated in increasing concentrations of alcohol (30, 50, 60, 70, 85, 95 and 100%) for 15 minutes each. The sample was stored in 100% alcohol, and was subjected to Critical point drying (CPD) and sputtered with gold ions and observed under the scanning electron microscope.

2.10 Cell morphology using histology

The scaffold containing cells was fixed in 10% formaldehyde solution and stored at 4 deg. C for 24 hours. The samples were then dehydrated and fixed in liquid paraffin. Thin sections were made (containing the sections of the scaffold) using a microtome and stained with eosin and hematoxilin stain and observed under an optical microscope.

2.11 Characterizing e- NOS gene expression using RT-PCR

Following the culture of ECs in 3D scaffold, RNA was extracted using standard protocols. It was quantified using a UV spectrophotometer. It was then reverse transcribed using the Qiagen RT kit to produce c-DNA. PCR was done using Qiagen PCR kit by using primers for the e-NOS gene

GAPDH primers:
reverse:5'- TCATGGATGACCTTGGCCAG-3'
forward:5'- GTCTTCACTCCATGGAGAAGG-3'
eNOS primers:
reverse:5' ATA GAA TTC ACCAGC ACC TTT GGG AAT GGC GAT-3'
forward:5'-ATA GAA TTC GGA TTC ACT GTC TGT GTT GCT GGA CTC CTT-3'

GADPH gene was used as the control. The PCR product was electrophoresed on 1% agarose gel and the bands were observed under UV light and analyzed using '*labworks*' software. The e-NOS/GADPH band intensity ratio was obtained in each case and compared. EC monolayer culture on polystyrene and on a coating of 3.5% gelatin + 1% genipin was used as controls.

3. RESULTS

3.1 Synthesis of gelatin scaffolds

Fig 2 shows the appearance of pure gelatin scaffolds as well as after cross-linking the scaffold with genipin. Pure gelatin scaffold is white in color, the cross linked scaffold is however blue due to the fact that genipin reacts with free amine groups in gelatin to impart a blue color to the scaffold. The surface of the scaffold in contact with the polystyrene mould had a thin layer of skin (about 0.5-1 mm).

(a) (b)

Fig 2: Gelatin scaffold- pure and cross- linked. (a) Pure gelatin 3.5%; (b) 3.5% gelatin cross linked with 1% genipin.

3.2 Scaffold morphology by optical microscopy

Fig 3 shows the morphology of 3 different concentrations of pure gelatin scaffolds, each freeze dried at -20 deg. C and -80 deg. C respectively. The scaffold morphology was similar in both horizontal and vertical cross- sections. The average pore sizes are summarized in table 1 below.

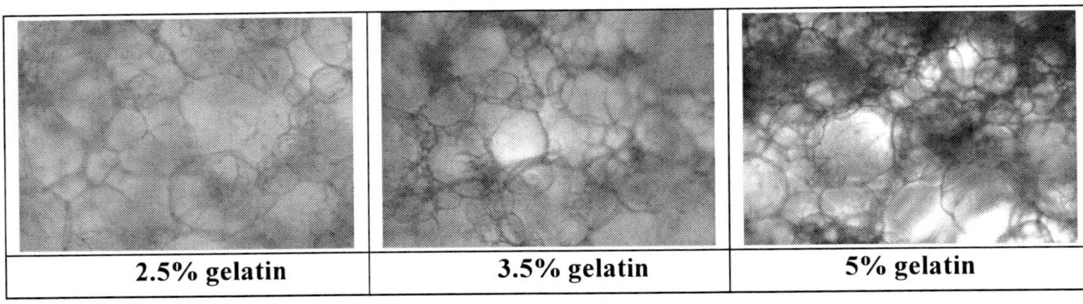

(a) Scaffolds freeze dried at – 20 deg. C

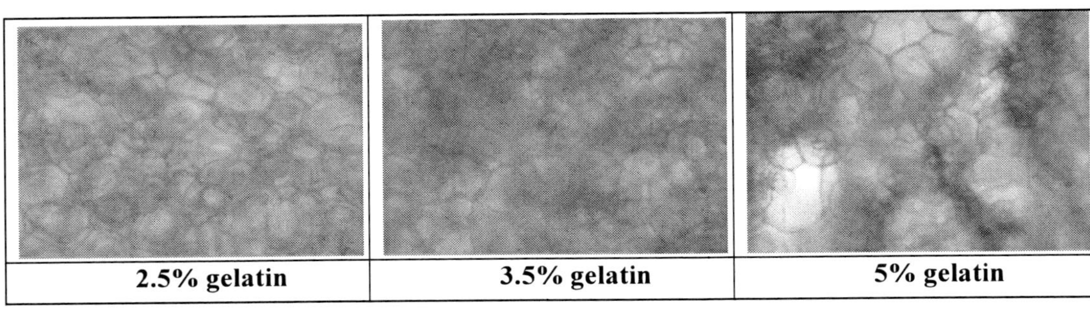

(b) Scaffolds freeze dried at - 80 deg. C

Figure 3: Optical micrographs showing the cross sections of pure gelatin scaffolds freeze dried at 2 different temperatures (-20 and - 80 deg. C)

Table 1: Pore size and distribution of 3 different concentrations of pure gelatin scaffolds freeze dried at 2 different temperatures

Freeze dry temp (deg. C)	Gelatin concentration	Average. pore size (μM)	Pore distribution
- 20	5%	150- 400	AsymmetricAsymmetric
	3.5%	100- 350	Homogenous
	2.5%	100- 350	
- 80	5%	50- 150	Asymmetric
	3.5%	50- 150	Homogenous
	2.5%	< 100	Homogenous

Fig. 4 shows the morphology of 3.5% pure as well as cross linked gelatin scaffolds as seen under an optical microscope- both in the dry condition and following its incubation in distilled water and in culture medium. As shown in the fig, the pores become more rounded after placing the scaffold in wet condition i.e., medium or water, but the pore doesn't collapse.

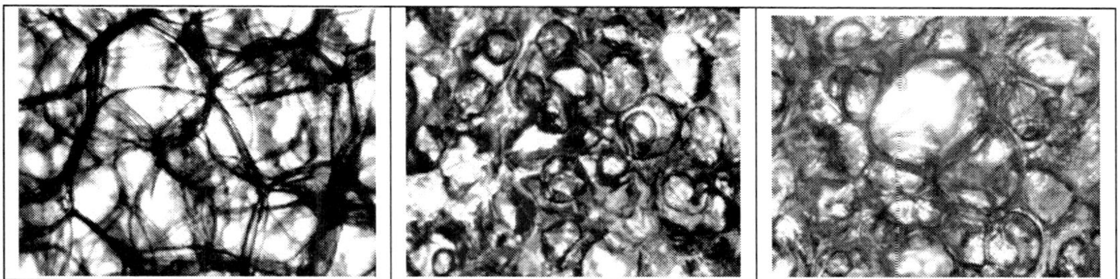

Dryin distilled waterin DMEM medium

(a) 3.5 % gelatin scaffold cross linked with 0.5% genipin

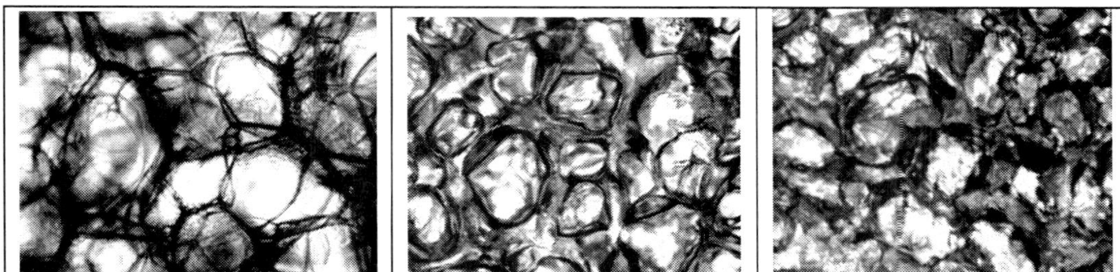

Dryin distilled waterin DMEM medium

(b) 3.5 % gelatin scaffold cross linked with 0.5% genipin

Figure 4: Optical micrographs showing the cross sections of 3.5 % gelatin scaffolds crosslinked with genipin- dry and wet conditions

3.3 Scaffold morphology by electron microscopy

Fig.5 shows the morphology of pure 3.5% gelatin scaffold as well as those that were cross linked with 0.5 and 1% genipin. The scaffold morphology was similar in both horizontal and vertical cross- sections. Table 2 indicates the average pore sizes of these scaffolds. As a general rule, scaffolds frozen at – 20 deg. C were asymmetric while those that were frozen at – 80 deg. C had a homogenous morphology. The cross- linking reaction did not have a significant effect on the pore size or the pore distribution.

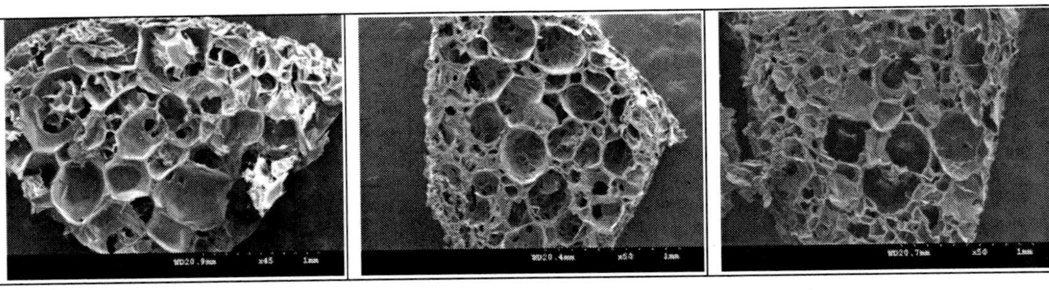

Pure 3.5 %gelatin gelatin + 0.5%genipingelatin + 1 % genipin
3.5% gelatin scaffolds freeze dried at – 20 deg. C

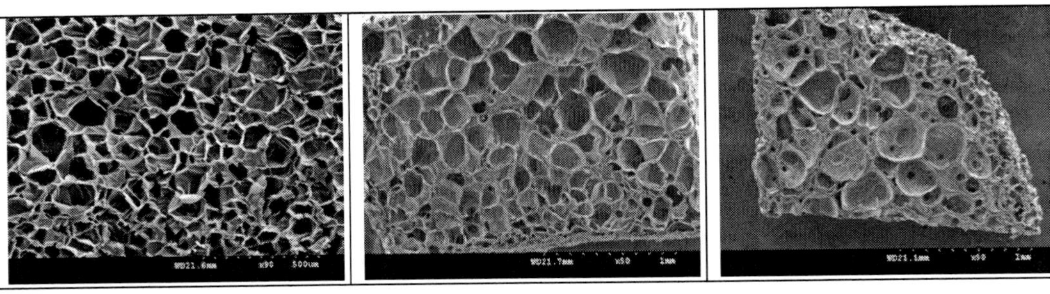

Pure 3.5 %gelatingelatin + 0.5%genipingelatin + 1 % genipin
3.5% gelatin scaffolds freeze dried at – 80 deg. C

Figure 5: Electron micrographs showing the cross sections of 3.5% gelatin scaffolds crosslinked with genipin

Table 2: Pore size and distribution of various 3.5% gelatin scaffolds freeze dried at 2 different temperatures - both pure and cross linked

Freeze dry temp (deg. C)	Genipin concentration	Average. pore size (μM)	Pore distribution
- 20	-	150- 400	Asymmetric
	0.5%	100- 350	Asymmetric
	1%	100- 350	Asymmetric
- 80	-	50- 150	Homogenous
	0.5%	50- 150	Homogenous
	1%	< 100	Homogenous

3.4 Swelling ratio

Fig 6 indicates the swelling ratio of pure as well as cross linked gelatin scaffolds as a function of time. The equilibrium swelling ratio of pure 5% gelatin scaffolds is about 18, while that of the 0.5 % genipin crosslinked scaffold is about 9 and that of 1% crosslinked scaffold is about 6.5. The crosslinked scaffolds reach their equilibrium swelling ratio in about 3 hours time while pure gelatin scaffold takes a much larger time (about 24 hours). Cross- liking the scaffolds therefore remarkably reduced the swelling ratio.

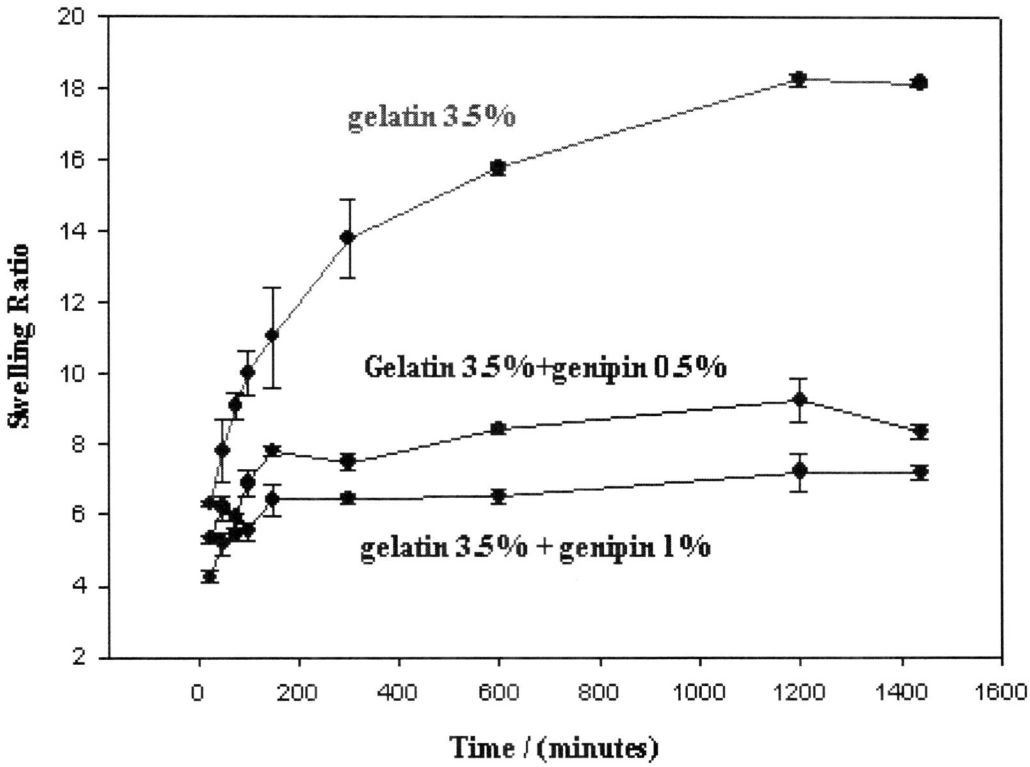

Figure 6: Swelling ratio of gelatin scaffolds

3.5 Characterizing degradation for gelatin scaffolds

The results of the degradation test using collagenase type II enzyme are summarized in fig 7. Pure gelatin scaffolds have a very high degradation rate- degrading completely within 3 hours. Cross linking the gelatin scaffolds with 0.5% genipin reduces the degradation rate, taking about 12 hours to completely degrade (data not shown) while cross-linking the scaffolds with 1% genipin slows down degradation rate even further with 40% of the scaffold remaining after 24 hrs (data not shown).

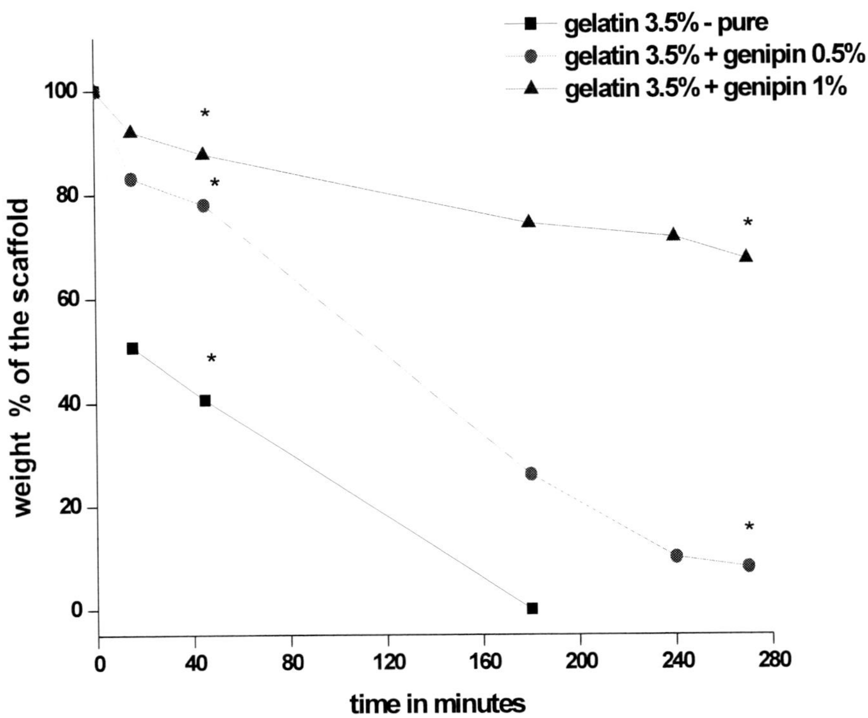

Figure 7: Degradation of gelatin scaffolds using collagenase II

3.6 Selecting a suitable scaffold for cell seeding

As shown in the fig 3, there was accumulation of excess gelatin in the 5% scaffolds. The 2.5% scaffold, on the other hand had a very small pore size and was very soft and difficult to handle. The 3.5% scaffold was therefore used for cross- linking reactions.

Table 3 showing the 2 different kinds of scaffolds used for EC seeding

3.5% GELATIN + 0.5% GENIPIN	FREEZE DRY TEMPERATURE	AVG. PORE SIZE	PORE DISTRIBUTION
Scaffold 1	- 20 deg. C	150- 400 μM	Asymmetric
Scaffold 2	- 80 deg. C	50- 100 μM	Homogenous

The 3.5% Gelatin scaffold cross- linked with 1% genipin was very rigid while the 0.5% cross- linked scaffold dissolved in culture medium within a few hours (fig 7). It was therefore decided to use 3.5% gelatin scaffold cross linked with 1% genipin with a recuced reaction time of 16 hours instead of 24 hours, for all the experiments. 2 different pore sizes of these scaffolds were used for the experiments as shown in table 3.

3.7 DNA assay for cell quantification

The results of the DNA assay to evaluate cell seeding into the scaffolds by methods 1 and 2 are shown in fig 8 and 9 respectively. In method 1, as shown in fig 8, there is a statistically significant increase in the cell number with time ($p<0.01$). The p-value for number of cells in each scaffold type is taken based on comparison with a similar scaffold type on a previous time point, for e.g., p- values of groups in day 3 are compared with those of day 1 and p-values of groups in day 5 are taken based on comparison with similar groups in day 3. There was no statistical difference between the cell numbers in scaffold 1 and scaffold 2 at any time point (day 1, 3 and 5). However, for cells seeded by method 2, scaffold 2 was more populated with cells than scaffold 1, at any given time point ($p<0.01$). On comparing the two different graphs, it is obvious that scaffolds seeded by method 2 contain much larger number of cells than those seeded by method 1. Based on the standard curve (not shown in fig), it was thus found that the seeding efficiency was about 30% for method 1 and about 50% for method 2. The seeding efficiency was calculated from the number of cells used for cell seeding and the number of cells left behind after the seeding procedure by either method 1 or 2. As shown in fig 12, the cell number in the gelatin scaffolds was saturated by day 3 in method 2. A reduction in cell number on day 5 may be due to the fact that cells had become overconfluent and some of the cells inside the confluent cell mass were no longer able to receive nutrients.

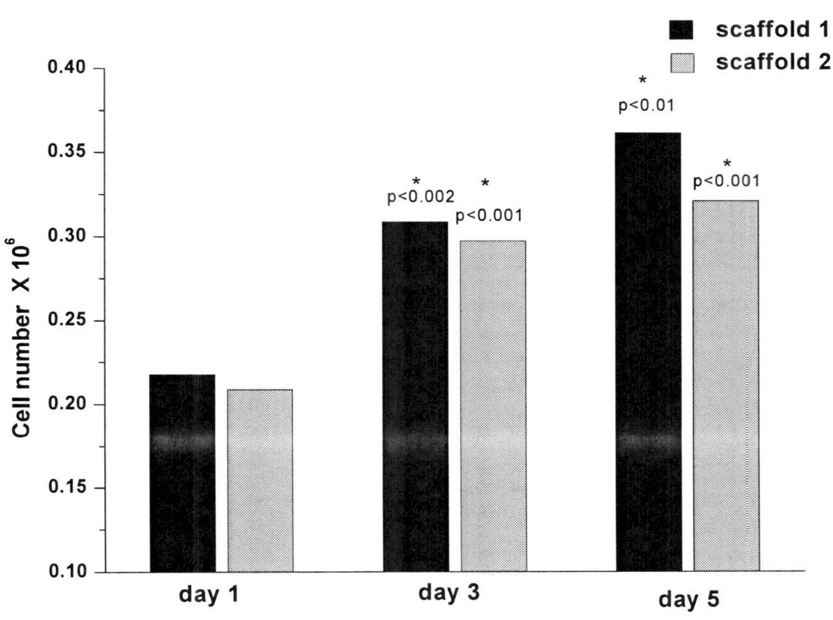

Figure 8: DNA assay after seeding by method 1

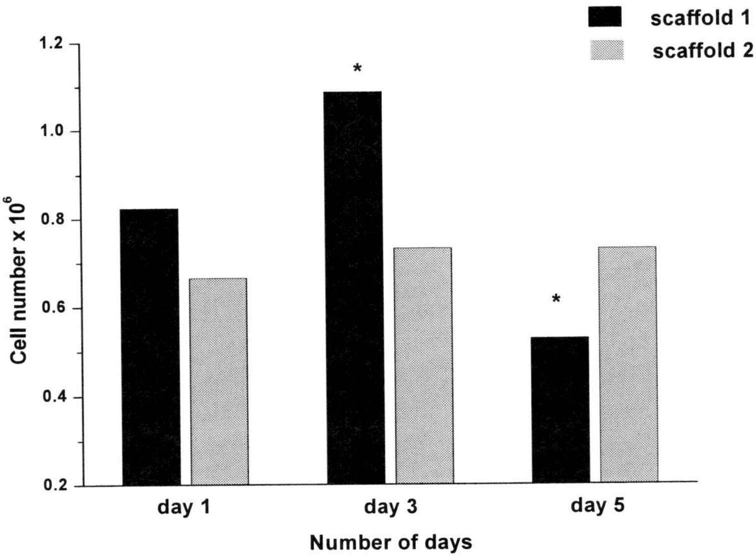

Figure 9: DNA assay after seeding by method 2

3.8 SEM of gelatin scaffolds containing cells

Fig 10 and 11 show the morphology of ECs seeded into scaffold 1 and scaffold 2 on the 1st, 3rd, 5th and 7th days respectively- both at lower and higher magnification. The morphology of the ECs in both scaffold types was found to be normal. An increase in the number of ECs can be appreciated in both scaffolds over a period of time. The ECs in scaffold 2 were of a more rounded morphology as compared to those of scaffold 1. However with increasing cell number, ECs in both scaffolds attained a rounded morphology. In general, ECs in scaffold 2 were less populated than scaffold 1.

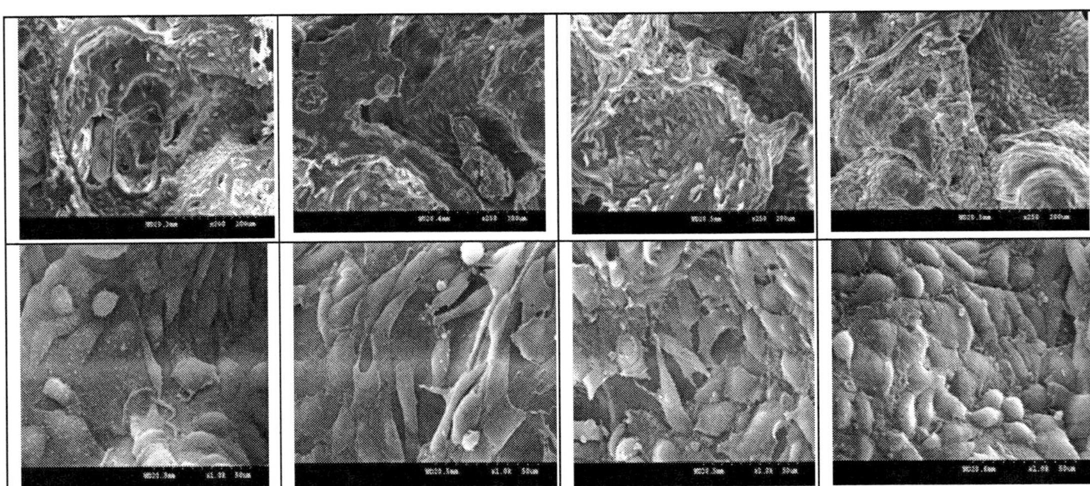

Day 1	Day 3	Day 5	Day 7

Figure 10: Electron micrographs of 'scaffold 1' containing cells shown in high magnification at the top and low magnification at the bottom

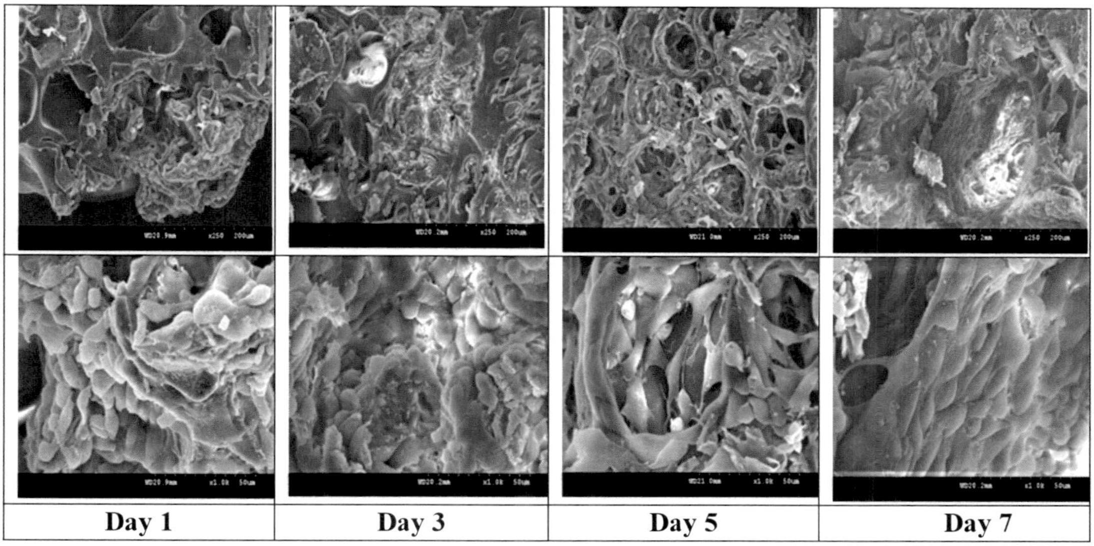

Figure 11: electron micrographs of 'scaffold 2' containing cells shown in high magnification at the top and low magnification at the bottom

3.9 Histology

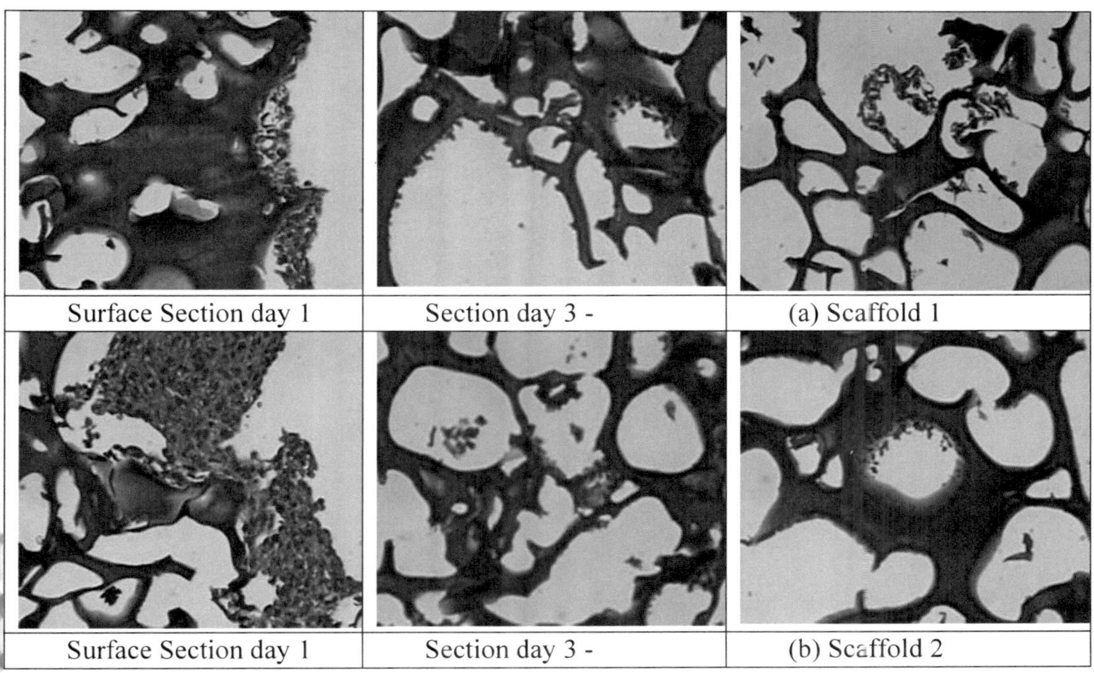

Figure 12: Histological sections of gelatin scaffolds seeded with endothelial cells

For both scaffold 1 and 2, there was an accumulation of cells on the surface as shown in fig 12. However, some cells migrated to the interior of the scaffold and had good cell viability.

3.10 Characterizing e- NOS gene expression using RT-PCR:

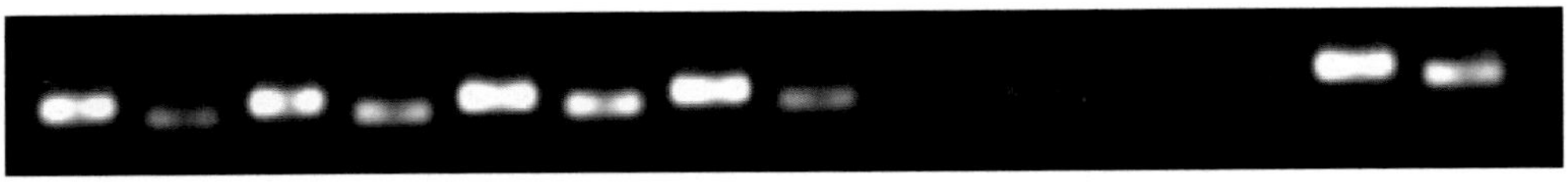

1a 1b 2a 2b 3a 3b 4a 4b 5a 5b 6a 6b 7a 7b

8a 8b 9a 9b 10a 10b 11a 11b 12a 12b

a - e-NOS geneb - GADPH gene
1, 2 – scaffold 1 (day 5); **3, 4** – scaffold 2 (day 5);**5, 7** – scaffold 1 (days 1, 3)
6, 8 - scaffold 2 (days 1, 3);**9, 10** – control 1 (days 3, 5);**11,12**- control 2 (days 3, 5)
Control 1- EC grown on tissue culture polystyrene (TCPS)
Control 2- EC grown on a 3.5% gelatin thin film crosslinked with 1% genipin

Figure 13: Electrophoretic gel picture showing RT-PCR data for e-NOS gene expression in ECs seeded into scaffolds – 1 and 2.

Gene expression of the nitric oxide synthase (e-NOS) gene in ECs cultured on both scaffold types was studied using RT-PCR at 3 different time points (days 1, 3 and 5). Two different controls were used. One was ECs cultured in a conventionl monolayer culture using TCPS and the other was ECs cultured on a 3.5% gelatin thin film crosslinked with 1% genipin. The results are shown in fig. 13 and 14. No e-NOS gene expression was seen in any of the samples on the 1st day (data not shown). On subsequent days, ECs in both kinds of scaffolds expressed e- NOS gene. While the controls in TCPS reached their maximum e-NOS expression on 3rd day itself, those in the 3D scaffolds, showed a gradual increase in e-NOS expression and reached their maximum in the 5th day that was comparable to the TCPS control. However, ECs cultured on the coating of 3.5% gelatin with 0.5% genipin didn't show any e- NOS expression. This may be attributed to the lack of cell attachment on the coating.

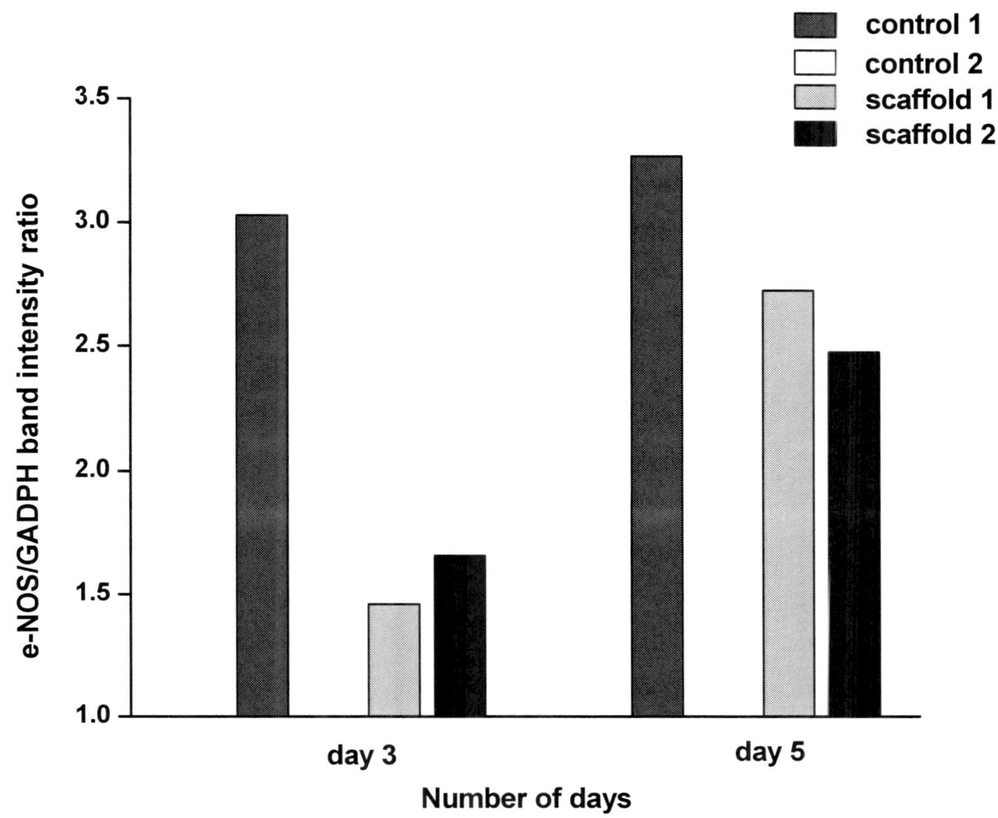

Figure 14: Comparing e-NOS gene expression of ECs in scaffold 1 and 2 with a conventional monolayer culture.

4. Discussion

Gelatin has been evaluated for use in skin, neural and even bone tissue engineering because of its biocompatibility and biodegradability [22,27]. Gelatin sponges have been recently tried for drug delivery. While gelatin has evaluated as a 3D scaffold to for tissue engineering applications, the cross- linking agents used for the purpose are mainly formaldehyde or glutaraldehyde, which have some toxicity to cells [22]. In this study, a natural cross- linked agent called genipin was used to make a novel 3D gelatin scaffold by the freeze dry method.

It was found that temperature had a significant effect on the scaffold morphology. For a given gelatin concentration, scaffolds frozen at -20 deg. C had much larger pore sizes than those frozen at -80 deg. C. This may be attributed to the increased number of nucleation sites at following rapid cooling to -80 deg. C than at -20 deg. C. The effect decreased with decreasing concentrations of gelatin, being prominent for 5% and 3.5% gelatin, while not being so prominent for 2.5% gelatin. Scaffolds frozen at – 80 deg C showed better homogeneity than those frozen at -20 deg. C.

Gooch et al., have stated that the pore size plays a very important role in modulating endothelial function [14]. It was therefore decided to use 2 different scaffold types of varying pore sizes (50-150 μM and 150- 400 μM) but same gelatin concentration for endothelial cell seeding. It was found that 5% concentration of gelatin was too high as it resulted in excess accumulation of gelatin within the scaffold, thus affecting the porosity of the scaffold. On the other hand, while 2.5% gelatin scaffold was homogenous, it was too soft and difficult to handle. Moreover, there was not as much variation in the pore size of 2.5% scaffolds whether frozen at -20 or -80 deg C. It was therefore decided to use 3.5% gelatin for the experiments with 2 different pore sizes as obtained by freezing at –20 and -80 deg Crespectively (50-150 uM and 150- 400 uM). In both cases, the porosity was found to be good in the vertical as well as horizontal sections with the smaller pore size scaffold showing homogenous distribution and the larger pore size scaffold showing a more asymmetric distribution.

Based on the results of the degradation test, it was found that gelatin scaffolds cross-linked with 1% genipin had maximum resistance to degradation and 1% genipin was therefore found to be ideal for cross linking; however, as the scaffolds obtained were very rigid, the reaction time for 1% genipin was shortened to 12-16 hrs instead of 24 hrs to reduce the brittleness of the scaffold.

Another problem faced during the synthesis of gelatin scaffolds by the freeze dry method was the formation of a skin layer on the surface of the scaffold in contact with the polystyrene mould. This could be due to the discrepancies in the transfer of heat on the top surface and the bottom surface in contact with the mould. In order to minimize the problem, the bottom surface was precooled in liquid nitrogen vapor before placing in the freeze dry equipment. While this reduced the thickness of the skin layer, it did not completely eliminate the problem.

After selecting the 2 different kinds of scaffold to be used for cell seeding, 2 different methods were tried for seeding the cells. One was a dropping method and the other was a shaking method. The shaking method was found to be much more efficient.

Vigorous shaking enabled better cell distribution into both kinds of scaffolds. Following cell seeding, it was found that scaffold 1 had a better seeding efficiency than 2. This may be attributed to the fact that the larger pore size of scaffold 1 resulted in easier distribution of ECs into the scaffold.

In general, ECs in scaffold 2 showed a more rounded morphology as compared to 1. However, once the scaffolds were well populated with cells, the ECs tended to be rounded in both types of scaffold, which might be attributed to a lack of space to spread. SEM as well as histological sections showed that many cells were confined to the surface of the scaffold and failed to reach the interior. This is understandable, especially in case of scaffold 2, where the smaller pore size can limit the migration of cells to the interior. But in either case, as shown by both histology and SEM, the cells that managed to reach the interior were healthy and viable and had a normal morphology. Gene expression studies with e- NOS gene, showed comparable expression with monolayer culture on tissue polystyrene, showing that ECs maintain their function well in these 3D scaffolds.

The ECs did not form any tube like structures or micro vessels and just continued to grow as a clump of cells even after 7 days of culture. Culture for longer time periods, addition of certain angiogenic factors and the co-culture of ECs with other cell types like SMC s can be pursued as strategies to form micro vessels with the scaffold.

5. Conclusion

In conclusion, gelatin scaffold cross linked with genipin was found to be a good biomaterial for the 3D culture of ECs. Scaffold 1 (pore size 150 – 400 μM) showed better seeding efficiency than scaffold 2 (pore size50 – 150 μM). Endothelial behavior was however found to be similar in both scaffold types and was comparable to its behavior in monolayer culture. While the ECs within these scaffolds didn't form any tubes in 7 days time, maybe larger culture periods could result in micro vessel formation. The scaffold also offers a good model to study the effect of different angiogenic factors on micro vessel formation within a biomaterial setting. The scaffold holds future promise for engineering some vascularised 3D tissue constructs.

B. Effect of Low Energy Laser Irradiation on Endothelial Cells

1. Introduction

Low energy laser irradiation (LELI) is being widely investigated and used as a form of medical therapy. It is the application of red and near infra-red light over injuries or lesions to improve wound and soft tissue healing and provide pain relief from both acute and chronic pain. LELI devices typically deliver energies between 5mW -1000 mW. LELI is used to increase the speed, quality and tensile strength of tissue repair, give pain relief, resolve inflammation and as an alternative to needles for acupuncture [28]. Investigators have noted an increase in the number of capillaries and the levels of angiogenic growth factors in patients who underwent Transmyocardial laser revascularization (TMR), a new form of therapy that improves symptoms in patients with stenotic coronary lesions [29,30].

It has been shown that the effect of LELI on the tissues is photochemical. Absorption of light by cytochromes and cell membranes followed subsequently by changes in the membrane permeability, intracellular calcium levels, singlet oxygen production and increased ATP production have all been proposed as possible mechanisms of laser action at the cellular level. Despite all these observations, there is no universally accepted theory as to the mechanisms of laser modulation of cellular processes [31].

Several studies using low-level continuous-wave visible and infrared laser irradiation have previously demonstrated inhibition or stimulation of cell growth, differentiation, motility, migration, and phagocytosis in vitro [32,33]. Steinlechner and Dyson have shown that keratinocyte and fibroblast proliferation can be stimulated directly and also indirectly through growth factors released from irradiated macrophages [34]. Studies on wound healing have shown that the duration of acute inflammation can be reduced by LELI with the proliferative phase of repair beginning earlier. The rate of wound contraction can be altered and angiogenesis increased [35,36]. A study on the effect of laser irradiation on the partially ruptured Achilles tendon in rats indicate that LELI improved cutaneous wound repair with an increase in angiogenesis and that the effect is a result of an inversely proportional relationship between wavelength and intensity [37]. A few studies have demonstrated that low-power laser irradiation increases production of vascular endothelial growth factor (VEGF) by smooth muscle cells (SMC), fibroblasts, and cardiac myocytes and stimulates endothelial cell (EC) growth in culture. Another study demonstrated that He-Ne (Helium Neon) laser irradiation (in doses of 60.5 J/cm^2) increased myocardial capillary permeability and the production of VEGF in myocardial micro vessels and in myocardium thus providing the experimental morphological evidence that myocardial microcirculation can be improved using He-Ne laser irradiation [38]. Recent work has focused on enhancing EC functions for TE applications (both TEBV and micro-vessels) by various kind of physical stimulation like cyclic mechanical stress [39-41], ultrasound etc [42,43].

Although a few studies have attempted to elucidate the effect of low energy laser irradiation (LELI) on ECs, they haven't been very comprehensive and have often given conflicting results. With recent evidences to indicate that the endothelial phenotype of the arteries and veins may well be phenotypically and genetically different [44], it is well possible that the effect of laser on these two cell types may also show variations. Then again, the nature of laser light can also influence the kind of effect on the cells. Most of the past studies have focused on venous ECs and have used (Helium Neon) He-Ne lasers. The diode laser is a much cheaper instrument than the He-Ne laser. My work therefore focused on the effect of LELI from a diode laser on bovine aortic arterial endothelial cells (BEC).

2. Materials and Methods

2.1 Endothelial cell culture

BEC (passage 15-25) were used for the experiments. The cells were cultured in H-DMEM (High glucose Dulbecco's Modified Eagle medium) from *Gibco Laboratories* containing L-Glutamine, pyridoxal hydrochloride, Sodium pyruvate, Sodium bicarbonate and supplemented with 10% FBS (fetal bovine serum) from *Biological Industries*. Penicillin-Streptomycin was used as antibiotic. BEC was incubated at 37 deg. C in a standard air/CO_2 incubator with humidified air containing 5% CO_2. For subculture, BEC were washed twice with PBS (phosphate buffered saline), trypsinised using 0.05% trypsin-EDTA from *Gibco laboratories* and seeded onto 24- well and 6- well plates for experiments.

2.2 Diode laser irradiation system

The diode laser system used for the experiments had an energy output ranging from 0-50 milliwatts (mW). Laser irradiation was transmitted on to the cell monolayer using optical fibers. Nature of the laser light was continuous and not pulsed. As shown in the figure below, during laser irradiation, the cells were kept in a standard air/CO_2 incubator at 37deg. C with humidified air containing 5% CO_2 (standard conditions).

Table 1: Different laser energies and exposure time periods used for the experiment.

Laser Energy (mW)	Exposure time (minutes)	Laser dosage (J/ cm^2)
25	10	3.75
	30	11.25
	60	22.5
	90	33.75
50	10	7.5
	30	22.5
	60	45
	90	67.5

Based on earlier experiments in our lab on the effects of He-Ne laser on human umbilical vein endothelial cells (HUVEC), the following two energies were used for the irradiation of endothelial cells for various time periods as indicated in the table below. Dosage was calculated as follows:

Power / Beam Area x Time = J/cm^2.

2.3 Cell viability assay using MTT

BEC (passage 15-25) were seeded in 24 well plates at a density of 5000 cells/ cm^2 and incubated under standard conditions for 24 hours. The endothelial cells were then treated with laser irradiation as per the scheme explained in table 1 above. Following laser irradiation, the cells were again incubated for a further period of 24 hours under standard conditions. Untreated BEC was used as control. The cell viability was then determined using the MTT test following standard protocols [45,46]. Briefly, following the incubation of BEC for 24 hours after laser treatment, the medium was aspirated and MTT solution (concentration 50 mg/ml) was added to each 24 well and incubated at 37 deg. C in the dark for 4-6 hours. The solution was then carefully aspirated. DMSO (dimethyl sulfoxide) was added to each well and again incubated at 37 deg. C for 15 min. The solution was then transferred to a 96 well plate and the absorbance was measured using an ELISA reader at 570 nm. The readings were taken in triplicates.

2.4 Cell cycle analysis using PI staining and FACS analysis

BEC (passage 15-25) were seeded in 6 well plates at a density of 30000 cells/ cm^2 and incubated under standard conditions for 24 hours. The endothelial cells were then treated with laser irradiation as per the scheme explained in table 1 above. Following laser irradiation, cells were again incubated for a further period of 24 hours under standard conditions. Untreated BEC was used as control. The cells were trypsinised and stained with Propidium Iodide following standard protocols [47,48]. The stained samples were then analyzed for their DNA content by a BD flowcytometer using *Cell Quest* software.

2.5 Expression of e-NOS gene using RT-PCR

BEC was cultured in 6 well plates at a density of 30000 cells/ cm^2 for 24 hours. The cells were treated with laser as per the experimental schema shown in table 1. Following laser irradiation, the cells were again incubated under standard conditions for 24 hours. RNA was extracted using standard protocols. It was then reverse transcribed using the Qiagen RT kit to produce c-DNA. PCR was done using Qiagen PCR kit using the following primers for the e-NOS gene:

GAPDH primers:
reverse:5'- TCATGGATGACCTTGGCCAG-3'
forward:5'- GTCTTCACTCCATGGAGAAGG-3'
eNOS primers:
reverse:5' ATA GAA TTC ACCAGC ACC TTT GGG AAT GGC GAT-3'
forward:5'-ATA GAA TTC GGA TTC ACT GTC TGT GTT GCT GGA CTC CTT-3'

GADPH gene was used as the control. The PCR product was electrophoresed on 1% agarose gel and the bands were observed under UV light and analysed using '*labworks*' software. The e-NOS/GADPH band intensity ratio was obtained in each case and compared.

3. RESULTS

3.1 viability test of laser irradiated BEC

The result of the cell viability assay was recorded as a histogram using *OriginPro 7.5* software and is shown in fig 2. BEC treated with both 25 and 50 mW laser irradiation for the specific time interval of 30 minutes showed a statistically significant increase ($p<0.05$)in cell viability as compared to non treated cells, as shown by the paired t- tests. At 10 and 90 minutes, the viability of the cells was comparable to that of control. However cells irradiated for a specific time interval of 60 minutes showed a decrease in viability for both energies. Treatment of cells beyond the 90 minute period had a negative effect on cell viability (data not shown). The cells did not show any morphological changes following the above experimental doses of laser irradiation (data not shown).

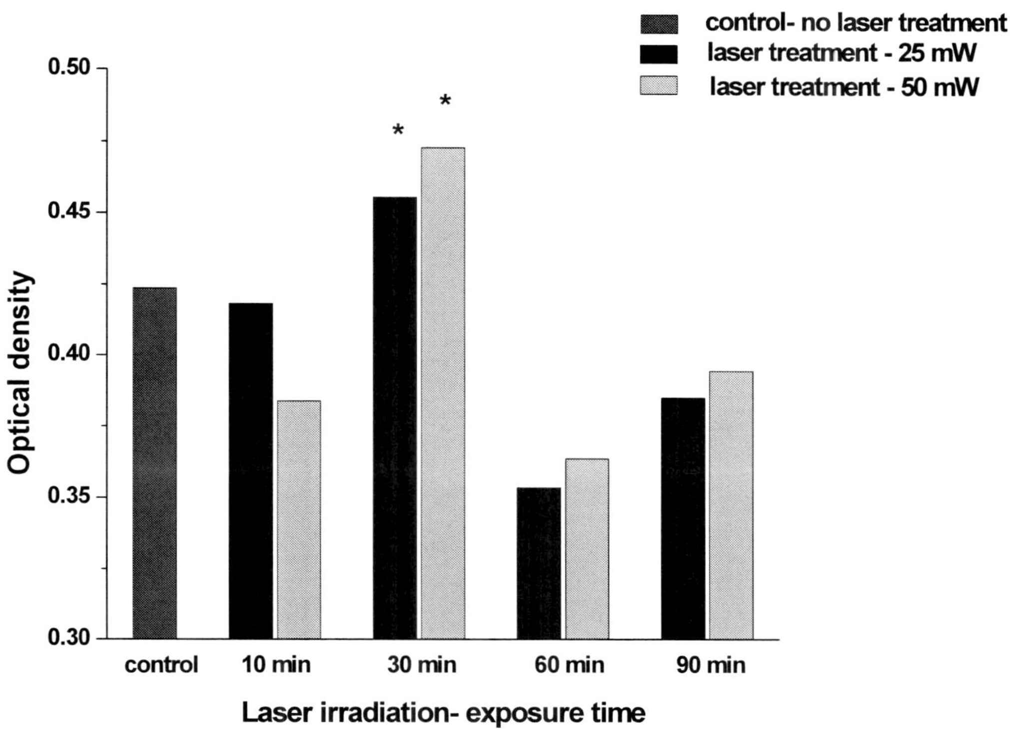

Figure 1: Results of the MTT assay for cell viability showing the effect of LELI on the viability of BEC

3.2 Cell cycle analysis of laser irradiated BEC

The above experiment was done to study if laser irradiation had any effect on the cell-cycle of BEC and therefore, to specifically confirm if the increase in cell viability was indeed due to an increase in the cell proliferation rate. The results are shown in fig 3, 4 and table 2. BEC treated with 25 mW laser irradiation for a 30 minute period showed a significant increase in the G_2/M population as compared to control (from 14% to 22%). No significant change was seen in the S-phase for cells treated with 25 mW laser energy. For cells treated with 50 mW laser energy, there was a slight increase in the S- phase population; however the G_2/M population did not show much change.

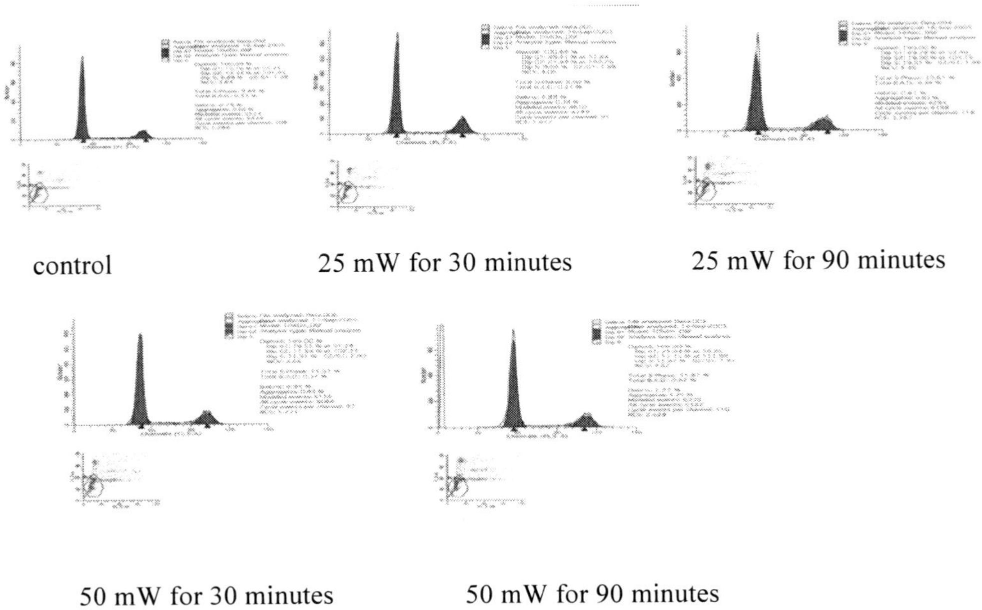

Figure 3: Cell cycle analysis

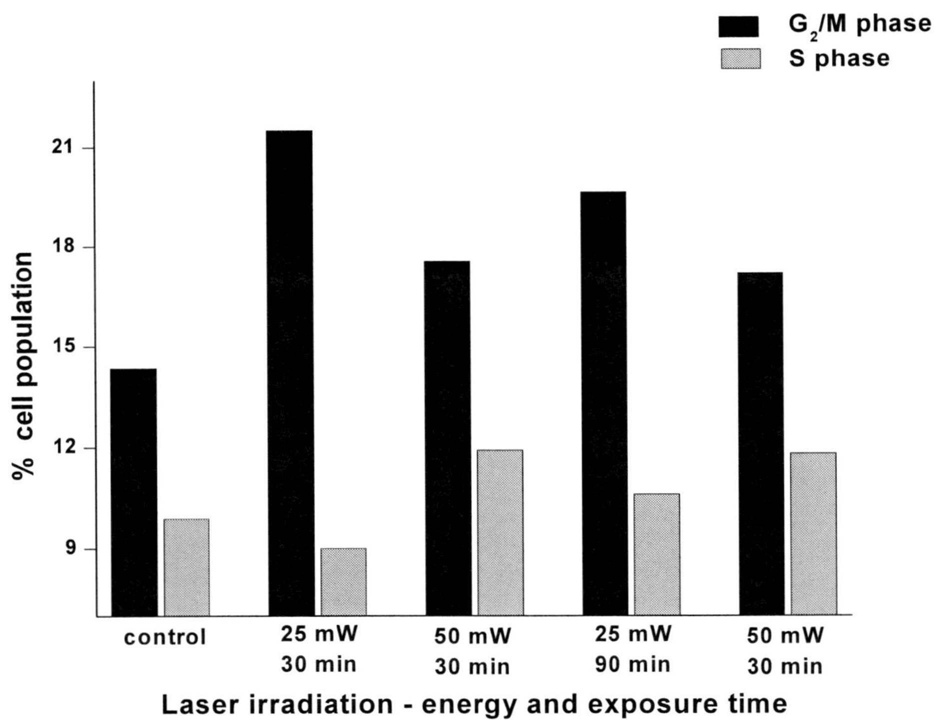

Fig 4: A comparative graph showing the effect of laser treatment on the cell cycle profile of BEC in different samples

Table 2: Cell cycle analysis results

Laser irradiation on BEC (energy in mW and exposure time in minutes)	Percentage of cell population in different stages of the cell cycle (%)		
	G_0/G_1phase	G_2/M phase	Sphase
control	75.76	14.34	9.89
25 mW for 30 min	69.51	21.49	9
50 mW for 30 min	70.55	17.54	11.91
25 mW for 90 min	69.79	19.60	10.61
50 mW for 90 min	71.04	17.15	11.61

3.3 e-NOS gene expression

The results of e-NOS gene expression for the various samples are shown in fig 5 and 6.

1- Control, **2, 4, 6, 8**- 25 mW for 10, 30, 60 and 90 minutes respectively

3, 5, 7, 9- 50 mW for 10, 30, 60 and 90 minutes respectively

Figure 5: RT-PCR results for e-NOS gene expression

Cells treated with 25 mW laser irradiation showed a statistically significant increase ($p<0.05$) in the expression of e-NOS gene with increase in time (up to 60 minutes), as shown by the paired t- tests. Treatment for longer duration of time (90 minutes) however resulted in a decrease in the e-NOS gene expression. Cells treated with 50 mW laser energy also showed a statistically significant increase in the expression of e-NOS gene with increase in the exposure time, with maximum exposure seen after 30 minutes and 90 minutes.

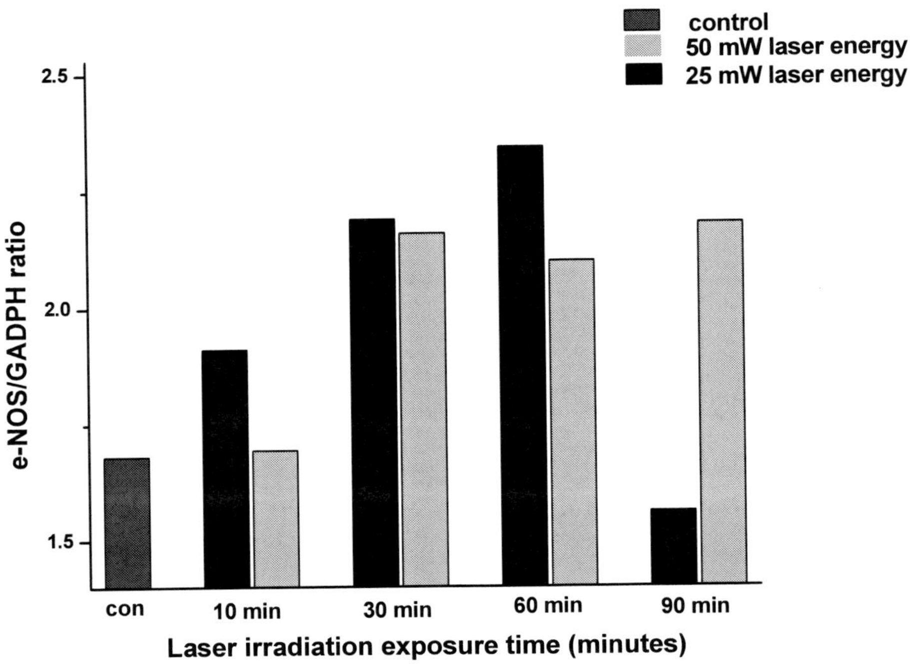

Figure 6: Results comparing e-NOS gene expression in various samples

4. Discussion

Atherosclerosis of the blood vessels leading to coronary artery disease and cardiac arrests constitutes a major health burden. Vascular grafts are being investigated to provide long-term relief in these diseases. While such grafts have shown reasonable degree of success for large vessels, replacement of the medium and small-sized vessels with such artificial grafts has been less successful, often rapidly leading to thrombosis and graft failure. It has been proposed that lining the grafts with a layer of endothelial cells would go a long way in providing a graft lining that is compatible with the patient blood and would consequently lead to better graft survival [49]. While in-vitro seeding of a graft with endothelial cells has not shown the kind of promise expected of them, there is still ongoing research to improve this technique. Common problems include lack of sufficient endothelial cells from the donor, their inability to withstand cyclical stresses etc.

Perhaps one way of addressing this problem is by enhancing the functions of the endothelial cells in-vitro, before implanting them onto the graft and into the patient. In the above work, BEC was treated with 2 different energies of laser irradiation (25 and 50 mW) for various time periods ranging from 10 to 90 minutes. It was found that treating the endothelial cells with a specific energy and dosage of LELI (25 mW for 30 minutes) leads to an increase in the proliferation rate. Exposure time therefore seemed to play a critical role on the viability of the BEC in addition to the irradiation energy.

The above finding was confirmed by analyzing the cell cycle profile of the treated cells, which showed that for this particular dosage there were more cells in the G_2/M phase and less in the G_0/G_1 phase as compared to controls. While a similar effect was also seen for cells irradiated with laser for 90 minutes (14% to 19%), though less prominently, this was not reflected in cell-viability studies with a corresponding increase in viability. This may be due to the fact that the 90 minute exposure time, while increasing the proliferation rate by acting on the cell-cycle pathway, may at the same time be decreasing the cell viability by some other mechanism. One reason why laser irradiation with 25 mW resulted in a specific increase in G_2/M population, with not much change in S- population may be because of the rapid progression of the cell cycle following irradiation with 25 mW and hence when the cell cycle profiling was done after 24 hours, most of the stimulated cells had moved past the S-phase into the G_2/M phase. However, for cells irradiated with 50 mW laser energy, the G_0/G_1->S->G_2/M transition might have been slower and so at the 24 hour period, most of the cells might still have been in the S- phase. Thus, both irradiation energy and time period seemed to play an important role on the cell- cycle profile of BEC.

With regard to e-NOS gene expression following laser irradiation, it was found that an exposure time of both 30 and 60 minutes resulted in a statistically significant increase in the gene expression. This was true for both 25 and 50 mW of laser energy used. The maximum increase in the expression of the e-NOS gene was seen with 25 mW laser irradiation for 60 minutes followed by 25 mW for 30 minutes.

Given that a lack of sufficient number of endothelial cells from a patient to line a vascular graft is a major factor limiting the success of such grafts, the above results might be of some benefit in increasing the cell number while reducing the culture time for the patient's own endothelial cells before being lined onto the graft. Results also showed that treating the ECs with 25 mW laser energy for 30 and 60 minutes led to an increase in e- NOS gene expression. Now, the e-NOS gene codes for *nitric oxide synthase* enzyme that in turn plays a pivotal role in nitric oxide (NO) production. Studies have shown that apart from its major vasodilatory function, NO also plays an important role in inhibiting platelet aggregation [50]. Thus a stimulated increase in NO production by the endothelial cells *in vitro* could be of potential benefit, if such cells were to be used in lining a vascular graft, thus inhibiting platelet aggregation *in vivo* as well, thereby leading to an increase in the survival of such vascular grafts *in vivo*. While an exposure of 25 mW laser energy for 60 minutes produced a maximum increase in e-NOS expression, the same exposure did not have a significant effect on cell proliferation rate. However, a 25 mW irradiation for 30 minutes produced a statistically significant increase in both the cell proliferation rate as well as e-NOS gene expression. Thus, overall, 25 mW laser irradiation for 30 minutes, having a dosage of 11.25 J/cm^2 was found to be most favorable in producing all the above beneficial effects, which could be potentially useful, if these irradiated cells could be used to line vascular grafts.

While these results are promising, a lot more work needs to be done in elucidating the signal pathways involved in the action of laser on these cells. *In vivo* experiments in animal models are essential to establish the validity of these *in vitro* results. Furthermore, tests could be done to study the effect of laser on EC-SMC interaction, as an abnormality in such interactions is of critical value in the pathogenesis of neointimal hyperplasia and progression of atherosclerosis.

5. Conclusion

The above work evaluated the effects of LELI from a diode laser on BEC. It was shown that, at a certain specific energy and dosage, the BEC proliferated at a more rapid rate. This was further confirmed by establishing that following such laser treatment, the cells moved from the resting phase to the synthetic and mitotic phases of the cell cycle. The same dosage also resulted in a significant increase in e- NOS gene expression. Our experiments showed that 25 mW laser irradiation for 30 minutes, having a dosage of 11.25 J/cm^2 was found to be most favorable in producing all the above beneficial effects in BEC. It is therefore postulated that ECs treated with specific laser energies for specified time periods could be of benefit in enhancing the properties of the endothelial cells used in lining a vascular graft, thereby increasing graft survival and reducing culture time. The above results, though preliminary could be further investigated for potential benefit in the development of medium and small vessel vascular grafts.

References

[1] Langer, R; Vacanti, JP. Tissue engineering. *Science 1993;260: 920–6.*

[2] Vacanti, JP. Beyond transplantation. Third annual Samuel Jason Mixter lecture. *Arch Surg 1988;123:545–9.*

[3] Vacanti, JP; Langer, R. Tissue engineering: the design and fabrication of living replacement devices for surgical reconstruction and transplantation. *Lancet 1999;354(Suppl 1): SI32–4.*

[4]Eiselt, P; Kim, BS; Chacko, B; et al. Development of technologies aiding large-Tissue engineering.*Biotechnol Prog* 1998; 14:134–40.

[5] Colton, C. Implantable biohybrid artificial organs. *Cell Transplantation* 1995; 4: 415–36.

[6] Folkman, J; Hochberg, M. Self-regulation of growth in three dimensions. *J Exp Med* 1973; 138: 745–53.

[7]Isner, JM; Asahara, T. Strategies for postnatal neovascularization. *J Clin Invest* 1999; 103: 1231–6.

[8]Folkman, J. Angiogenesis in cancer, vascular, rheumatoid and other disease. *Nature Med* 1995; 1: 27–31.

[9] Norrby, K. Angiogenesis: new aspects relating to its initiation and control. *APMIS* 1997; 105: 417–37.

[10]Zagzag, D. Angiogenic growth factors in neural embryogenesis and neoplasia. *Am J Pathol* 1995; 146: 293–309.

[11] Cassell, OC; Hofer, SO; Morrison, WA; Knight, KR. Vascularisation of tissue-Engineered grafts: the regulation of angiogenesis in reconstructive surgery and in disease states. *Br J Plast Surg* 2002; 55:603–10.

[12] Salacinski, HJ; Goldner, S; Giudiceandrea, A; Hamilton, G; Seifalian, AM; Edwards, A; Carson, RJ. The mechanical behavior of vascular grafts: a review. *J Biomater Appl* 2001;15: 241–78.

[13]Sieminski, AL; Gooch, KJ. Biomaterial microvasculature interactions. *Biomaterials* 21 (2000) 2233-2241.

[14] Peters, MC; Polverini, PJ; Mooney, DJ. Engineering vascular networks inporous polymer matrices. *J Biomed Mater Res* 60: 668–678, 2002.

[15] Xiao et al. Tissue-engineered micro vessels on three-dimensional biodegradablescaffoldsusing human endothelial progenitor cells. *Am J Physiol Heart Circ Physiol* 287: H480–H487, 2004;

[16] John, P; Courts, A; Ward, AG; Courts (Ed.). The Science and Technology of Gelatin. *Academic Press, New York*, 1977, p. 138.

[17] Braunwald, NS; Gay, W; Tatooles, CJ. Evalaution of crosslinked gelatin as a tissue adhesive and hemostatic agent:An experimental study. *Surgery* 59 (1966) 1024.

[18] Michon, C; Cuvelier, G; Relkin, P; Launay, B. *Int. J. Biol. Macromol.* 20 (1997) 259.

[19] Sung, HW; Huang, RN; Huang, LLH; Tsai, CC; Chiu, CT. Biocompatibility study of a biological tissue fixed with a naturally occurring crosslinking reagent. *J.Biomed. Mater. Res.* 42 (1998) 560.

[20] Huang, LL; Sung, HW; Tsai, CC; Huang, DM. *J. Biomed. Materials. Res.* 42 (1998) 568

[21] Sung, HW; Huang, DM; Chang, WH; Huang, LL; Tsai CC; Liang IL. Gelatin- derived bioadhesives for closing skin wounds: an in vivo study. *J Biomater Sci Polym Ed.* 1999;10(7):751-71.

[22] Yao, CH; Liu, BS; Chang, CJ; Hsu, SH; Chen, YS. Preparation of networks ofgelatin and genipin as degradable biomaterials. *Materials Chemistry and Physics* 83 (2004) 204–208

[23] Kang, HW; Tabata, Y; Ikada, Y. Fabrication of porous gelatin sca!olds for tissue engineering. *Biomaterials 20* (1999) 1339-1344.

[24] Choi, YS; Hong, SR; Lee, YM;Song, KW; Park, MH; Nam, YS. Study on gelatin-containing artificial skin: I. Preparation and characteristics of novel gelatin-alginate sponge. Biomaterials 20 (1999) 409-417.

[25] Choi, YS; Hong, SR; Lee, YM;Song, KW; Park, MH; Nam, YS. Studies on Gelatin-Containing Artificial Skin: II. Preparation and Characterization of Crosslinked Gelatin-Hyaluronate Sponge. *J Biomed Mater Res (Appl Biomater)* 48: 631–639, 1999.

[26] Kim, YJ; Sah, RL; Doong, JY; Grodzinsky, AJ. Fluorometric assay of DNA in cartilage explants using Hoeschst 33258. *Anal Biochem* 1988; 174:168-76.

[27] Kawai, K; Suzuki, S; Tabata, Y; Ikada, Y; Nishimura, Y. Accelerated tissue regeneration through incorporation of basic fibroblast growth factor-impregnated gelatin microspheres into arti"cial dermis. *Biomaterials 21* (2000) 489-499.

[28] Epub 2003 Mar 07.*Proc National Academy of Science U S A. 2003 Mar18; 100(6): 3439-44*

[29] Allen, KB; et al. Holmium:YAG laser system for transmyocardial revascularization*Expert Rev Med Devices*. 2006 Mar; 3(2):137-46

[30] Lamy, A; et al. Transmyocardial laser revascularization may relieve angina and improve myocardial perfusion among patients with angina refractory to medical treatment.*Evid Based Cardiovascular Med.* 1997 Sep; 1(3):77

[31] Uzdensky, AB. Isolated neuron response to blue laser.phenomenoloy and possible mechanism. *From URL http://www.photobiology.com/v1/uzdensky2/default.htm*

[32]Kipshidze, N; Keelan, MM; Horn, JB; et al. Biomodulation of vascular cells with low-power red laser light in vitro. *Int J Cardiovasc Med Sci.* 1998: 1: 241–245

[33] VanBreugel, HH; Bär, PR. Power density and exposure time of He-Ne laser irradiation is more important than total energy dose in photo-biomodulation of human fibroblasts in vitro. *Lasers Surg Med.* 1992; 12: 528–537

[34] Steinlechner C, Dyson M: The effect of low level laser therapy on the proliferation of keratinocytes. *Low Level Laser Therapy;* 1993; 5(2): 65.

[35] Dyson, M; Young S. Effect of Laser Therapy on Wound Contraction and Cellularity in Mice. *Lasers in Medical Science.* 1986; 1: 125.

[36] Dyson, M. Cellular and subcellular aspects of low level laser therapy. *Progress in low level laser therapy.* Eds. T. Ohshiro and R.G. Calderhead, John Wiley & Sons, England. 1991, p. 221

[37] Salate, AC; Barbosa, G. Effect of In-Ga-Al-P Diode Laser Irradiation on Angiogenesis in Partial Ruptures of Achilles tendon in Rats. *Photomedicine and Laser Surgery* Oct 2005, Vol. 23, No. 5: 470-475

[38] Wei-guang, Z; Chang-yan, W; Wen-xiao, P; Long, T; Jia-liu, X.Low-power Helium-Neon laser irradiation enhances the expression of VEGF in murine myocardium *Chin Med J* 2004; 117(10):1476-1480

[39] Wedgwood, S; Bekker, JM; Black, SM. Shear stress regulation of endothelial NOS in fetal pulmonary arterial endothelial cells involves PKC.*Am J Physiol Cell Physiol* 272: C421-C427, 1997.

[40] Howard, AB; Alexander, RW; Nerem, RM; Griendling, KK; Taylor, WR. Cyclic strain induces an oxidative stress in endothelial cells. *Am JPhysiol Cell Physiol* 272: C421-C427, 1997.

[41] Fujiwara, K. Mechanical stresses keep endothelial cells healthy: beneficial effects of a physiological level of cyclic stretch on endothelial barrier function. *Am J Physiol Lung Cell Mol Physiol* 285: L782-L784, 2003; doi:10.1152/ajplung.00142.2003

[42] Raz, D; Zaretsky,U; Einav, S; Elad, D. Cellular alterations incultured endothelial Cells under therapeutic ultrasound irradiation. *2003 Summer Bioengineering Conference, June 25-29, Sonesta Beach Resort inKey Biscayne, Florida*

[43] Karsch, U; Henn, W, Seyfert, UT; Steudel, WI; Reif, J. Effect of high powerultrasound on endothelial cells - an invitro study of the endothelium - hemostasis interaction. *Clinical Hemorheology and Microcirculation* Volume 27, Number 2 / 2002, 123-135

[44] Sato, TN. Emerging concept in angiogenesis: specification of arterial and venous endothelial Cells. *British Journal of Pharmacology* (2003) 140, 611−613.

[45] van de Loosdrecht, AA; et al. *J. Immunol. Methods* 174: 311-320, 1994.

[46] Ferrari, M; et al. *J. Immunol. Methods* 131: 165-172, 1990.

[47] Crissman, HA; Steinkamp, JA. Rapid simultaneous measurement of DNA, protein and cell volume in single cells from large mammalian cell populations. *J. Cell Biol., 59:766, 1973.*

[48] Krishan A. Rapid flow cytofluorometric analysis of cell cycle by propidium iodide staining. *J. Cell Biol., 66:188, 1975.*

[49] Kakisis, JD; Liapis, CD; Breuer, C; Sumpio, BE. Artificial blood vessel: The Holy Grail of peripheral vascular surgery. *J VascSurg 2005; 41:349-54*

[50] Klinge, JM; Topf, HG; Trusen, B; Rauh, M; Rascher, W; Dotsch, J. Endothelial cells play an important role in the antiaggregatory effect of nitric oxide. *Crit Care Med. 2003 Jul; 31(7):2010-4.*

In: Biomaterials Research Advances
Editor: J. B. Kendall, pp. 67-91
ISBN: 978-1-60021-892-7

Chapter 4

RESORBABLE POLYMERS IN SPINAL SURGERY

T. U. Jiya, T. H. Smit and P. I. J. M. Wuisman
Department of Orthopaedic Surgery VU Medical Centre, Amsterdam
And Skeletal Tissue Engineering Group, Amsterdam

ABSTRACT

The potential utility of polymer based resorbable implants in structural support applications, as biological container, and protective adhesion barrier in spinal surgery has been the focus of research in recent times. Accumulated preclinical experience coupled with improved polymer chemistry has allowed the clinical introduction of these devices into spinal surgery. The main focus of research has been on implants intended to aid spinal fusion (cages, screws, rods, scaffold carriers) and protective adhesion barriers for the neural elements.

Resorbable fusion implants are manufactured from polymers of which polylactic acid (PLA) is the most significant component. Preclinical studies have demonstrated adequate biocompatibility, sufficient stiffness and strength whilst load transfer to healing graft bone is gradual. Preclinical studies have also indicated that several parameters, including crystallinity, molecular weight, implant design and method of sterilization, which all affect the physical and biological properties of PLA based implants, may influence their clinical performance. The most clinical experience have been gained using the co-polymer 70:30 PLDLLA cage implant with which 87-97% spinal fusion rates have been reported, rates comparable to that seen with routinely applied non-resorbable implants. Also a novel application of a resorbable film as a protective adhesion barrier to neural elements has been reported with promising results.

The clinical success of various PLA based implants in spinal surgery seems to be influenced by the anatomical site and the corresponding local physiological environment. The mode of failure of PLA based implants is time dependent and is influenced by the nature of static and dynamic loading in vivo. Consequently future research should be directed towards models that will help understand and predict the biological and biomechanical behavior and performance of these polymers. Furthermore, there is need to clarify the influence of anatomical site of implantation on the behavior of the implant, with a goal towards the development of site specific implants.

INTRODUCTION

The use of resorbable polymers for surgical implants has been propagated since the mid 60's.(Kulkarni et al., 1966) Although initially as resorbable sutures, the past 2 decades has witnessed the successful employment of resorbable polymers in fracture fixation and reconstruction of craniofacial defects.(Middleton & Tipton, 2000) The most commonly used resorbable polymers for orthopaedic implants are polyglycolic acid (PGA), polylactic acid (PLA), and PLA-PGA co-polymers.(Middleton & Tipton, 2000) More recently, the use of resorbable polymers has expanded to the realms of spinal reconstructive surgery.

The inherent limitations of the current metal and carbon fiber based spinal implants generated the impulse for the evolution of resorbable polymer implants. Metallic implants promote stress shielding, interfere with radiological evaluation of fusion, and are inadequate for use as protective device for neural elements.(Blumenthal & Ohnmeiss, 2003; Cizek & Boyd, 2000; Cook et al., 2004; Elias et al., 2000; Leclercq, 2001; McAfee et al., 1999; Naderi & Arda, 2001; Santos et al., 2003; Togawa et al., 2003; Togawa et al., 2004; Tullberg, 1998) Furthermore, late implant related complications, including allergy to these metals and associated risk of re-intervention for implant removal, does exist. Tullberg reported on the failure of a carbon fiber cage implant, demonstrating that cage breakage can occur in the presence of a nonunion.(Tullberg, 1998) Particle debris was observed as black tissue surrounding the dura and nerves. Other retrieval studies have confirmed the shortcomings of traditional non-resorbable cage implants. Togawa et al. reported on 78 cages that clinically failed and were retrieved from 48 patients.(Togawa et al., 2004) There were 8 carbon fiber reinforced cages and 70 metallic cages. They saw no direct bone apposition to either carbon fiber or metal in any cage. At least a few particles of debris were present, often within macrophages, in more than two-thirds of the cages. Resorbable polymers offer a valuable alternative to these drawbacks and consequently, their potential utility in structural support applications, as biological container and protective adhesion barrier in spinal surgery has stimulated intense research work in the last decade.

Poly-lactic acid (PLA), the most common aliphatic polyester polymer used in manufacturing spinal implants, can be tailored to meet the physiological and biomechanical requirements of the human spine. Biocompatibility is generally considered to be adequate even with respect to neural elements, and degradation products are completely eliminated without apparent adverse inflammatory tissue reaction.(de Medinaceli et al., 1995; Gautier et al., 1998; Gogolewski et al., 1993; Gogolewski, 2000; Matsumoto et al., 2002; Oudega et al., 2001; van Dijk et al., 2002d; van Dijk et al., 2005; Bergsma et al., 1995b) There appears to be no significant immunogenic or carcinogenic risk associated with the use of resorbable PLA and PGA based implants. Resorbable fusion devices have been shown to provide adequate stiffness and strength whilst load transfer to healing graft bone is gradual minimizing stress shielding.(Cahill et al., 2003; Cornwall et al., 2004; Hojo et al., 2005; Pflugmacher et al., 2004; Shikinami & Okuno, 2003; Soderlund et al., 2004; van Dijk et al, 2005; Wuisman et al., 2002} Resorbable protective film barriers on the other hand prevent adhesions to neural tissues whereas tubular devices promote uncomplicated healing and regeneration of the neural elements.(Klopp et al., 2004; Welch et al., 2002) Current applications of resorbable polymers in spinal surgery includes interbody fusion devices, posterior spinal graft containment, anterior cervical and lumbar spine tension band plating, posterior trans-laminar and trans-pedicular instrumentation, protective sheath for neural elements, and bone graft harvest site reconstruction. Also the potential utility of resorbable polymer based tubular scaffolds for cell transplantation and as guidance channels for regenerating axons in the injured spinal cord has been reported.(Oudega et al., 2001; Patist et al., 2004)

The field of spinal surgery has provided the breeding ground for one of the most recent advancements in the clinical application of resorbable polymer technology and this chapter is intended to give a critical overview of the currently available research data on this subject. We will conclude by discussing future directions in spine related resorbable technology.

PHYSICAL PROPERTIES

PLA and PGA are high molecular weight polyhydroxyacids formed by a ring opening polymerization of cyclic di-esters.(An et al., 2000) The carbon back bone is hydrolytically unstable such that when placed in an aqueous medium, the chains will degrade over time. PLA exists as two optical isomers, the naturally occurring Poly-L-lactide (PLLA) and poly-D-lactide (PDLA). Lactobacillus and E. Coli strains can be genetically engineered to selectively produce D-isomers or L-isomers.(Zhou et al., 2003; Bendix, 1998) Both PLLA and PDLA are semi crystalline whilst a mixture of the two is amorphous. When both isomers are mixed in equal portions (50:50) an entirely amorphous racemic polylactide is produced. Non-racemic mixture of the isomers particularly in the ratio 70:30 PLDLLA (contains 70% L-isomer and 30% racemic polylactide) is currently one of the most frequently used raw material in the fabrication of spinal implants.

The physical properties of resorbable polymers are defined by the crystallinity, average molecular weight, molecular weight distribution, melting point, glass transition temperature and impurities.(Pietrzak et al., 1996; Pietrzak et al., 1997; Weir et al., 2004c; Weir et al., 2004a; Weir et al., 2004b; Wuisman & Smit, 2006; Gogolewski et al., 1996a; Gogolewski & Mainil-Varlet, 1996b; Gogolewski & Mainil-Varlet, 1997) With all other parameters constant, a polymer with a higher crystallinity will be stronger, stiffer and typically degrade at a slower rate. Polymers like PLLA crystallize easily due to their symmetrical repeat unit and flexibility of the long chain. High molecular weight crystalline PLLA however is hard and brittle, a shortcoming which restricts its application in pure form. For this reason blending with racemic PLDLA or other polymers such as propylene oxide and polyethylene oxide is sometimes done to enhance its mechanical properties.

In addition to the crystallinity the molecular weight also has a major impact on the properties and the degradation kinetics of resorbable polymers.(Mainil-Varlet et al., 1997b; Mainil-Varlet et al., 1997a; Wuisman & Smit, 2006) Polymer strength increases with molecular weight by formation of secondary bonds between the chains and thereby slowing down the rate of degradation. The process of fabrication, annealing and sterilization of PLLA can significantly alter its properties of crystallinity and molecular weight. Weir et al investigated the processing of PLLA by compression molding (into plates) and extrusion (into rods).(Weir et al., 2004b) They reported that the molecular weight of PLLA decreased through both processing routes, and subsequent annealing at 120 degrees centigrade for 4 hrs in a preheated air circulating oven and sterilization by EtO.

The glass transition temperature is another important but not independent factor that reflects the physical characteristics of resorbable polymers. Glass transition temperature of polylactides is typically in the range of 55-60 degrees.(Wuisman & Smit, 2006) This is important for resorbable implants that need to maintain adequate mechanical integrity at local conditions that may be well above body temperature for a period long enough to acquire fusion. When under constant or dynamic loading for a long period of time, resorbable polymers tend to creep with resultant implant deformation and micro-fracturing.(Krijnen et al., 2004; Wuisman & Smit, 2006) This effect is more pronounced with polymers that have a lower glass transition temperature.

Other factors that help determine the physical properties of resorbable polymers are the presence of impurities (such as residual manomers, water and free radicals) and handling of the finished product particularly with respect to sterilization and packaging.(Bostman & Pihlajamaki, 2000a; Bostman & Pihlajamaki, 2000b; Wuisman & Smit, 2006) Sterilization can have adverse effects on the physical properties of resorbable polymer implants.(Loo et al., 2005a; Loo et al., 2005b; Loo et al., 2006; Matthews et al., 1989; Nuutinen et al., 2002; Wuisman & Smit, 2006) Hospital steam sterilization techniques for instance use high moisture content with temperatures well in excess of 100 degrees, thereby exceeding the glass transition temperature of PLA and PGA implants and is therefore highly unsuitable for the sterilization of these implants. Ionizing radiation on the other hand can inflict severe damage in polymer implant with adverse effects on its physical, biological and biomechanical properties.(Loo et al., 2005a; Nuutinen et al., 2002) Chemical sterilization tends to leave residues in harmful quantities on the surface of the implant which in turn can be reduced to safe margins by degassing or aeration. Consequently, chemical sterilization of PLLA using ethylene oxide does not adversely affect its intrinsic viscosity, crystallinity and mechanical properties.(Matthews et al., 1989; Nuutinen et al., 2002) Plasma sterilization likewise appears to have little effect on material properties and in particular no detrimental effect with respect to changes in the molecular weight and tensile strength of PLA.(Gogolewski et al., 1996a; Nuutinen et al., 2002; Wuisman et al., 2006)

Several factors help define the physical properties of resorbable polymers, of which the most clinically relevant aspects have been discussed above. Literature is sometimes misleading because authors do not sufficiently describe the properties of the polymer implants used, making comparison between different studies difficult if not impossible. Authors of future reports should therefore make an effort to give data on the physical properties as well as the sterilization methods of the polymer used.

Biocompactibility and Degradation

In general PLA and PGA based implants have demonstrated a high level of biocompatibility and particularly PLLA has been shown to be highly compatible to bone and neural tissues.(An et al., 2000; Bergsma et al., 1995b; Bostman & Pihlajamaki, 2000a; de Medinaceli et al., 1995; Mainil-Varlet et al., 1997b; van Dijk et al., 2002d; van Dijk et al., 2005; Wuisman & Smit, 2006) Matsuesue et al. implanted PLLA rods into the intramedullary cavity of the femur of rabbits and found no signs of toxicity or adverse tissue reactions to the material upon complete resorption of the rods.(Matsusue et al., 1995) In a recent long term study by the same group, both electron microscopy and histology demonstrated excellent biocompatibility and resorbability of a hydroxyapatite PLLA composite in rabbits over a period of 5-7 years.(Hasegawa et al., 2006) Matsumoto et al. implanted plates made of pure PLLA and a hydroxyapatite PLLA composite in the epidural space of white Japanese rabbits and demonstrated the safety of both plates to spinal neural elements.(Matsumoto et al., 2002)

Biocompatibility of PLA and PGA based resorbable implants has also been documented in the clinical setting. In a recent long term study by Coe and Vaccaro cages fabricated from a 70:30 ratio of poly (L-lactide-co-D,L-lactide) copolymer were implanted in 31 patients after being impacted with autologous iliac crest graft.(Coe & Vaccaro, 2004) At a minimum follow up of 2 years, significantly exceeding the biological life expectancy of the implant, no patient exhibited any signs of adverse tissue reaction or regional osteolysis. Comparably favorable results have been reported in several other animal and clinical studies.(Alexander et al., 2002; Austin et al., 2003; Cahill et al., 2003; Coe, 2004; Couture & Branch, 2004; Lanman & Hopkins, 2004; Lanman & Hopkins, 2004; Vaccaro et al., 2004; van Dijk et al., 2002a; van Dijk et al., 2002b; van Dijk et al., 2005) Yet historical data suggests that clinically significant adverse tissue reaction can result from resobable PLA and PGA implants.(Bergsma et al., 1995a; Bostman & Pihlajamaki, 1998; Bostman & Pihlajamaki, 2000b; Bostman et al., 2005) Bostman and Pihlajamaki reported on 2528 patients operated on using implants made of PLA and PGA in which 108 (4.3%) patients developed clinically significant local inflammatory sterile tissue reaction. However only one of the 108 patients received a PLA implant whereas the remaining 107 had a PGA implant inserted. Also other studies involving spinal neural tissues have shown particularly PGA based implants to adhere to and become inseparable from the dura matter only 2 weeks after implantation.(Gautier et al., 1998) This effect was not observed with PLA based implants. Although this and several other studies have indicated the superior biocompatibility properties of PLA relative to PGA, reports by some authors regarding massive osteolysis and subsidence seen after PLIF procedures with PLDLA cages remains worrisome.(Herceg et al., 2004) The extent and impact of such potentially deleterious effects as illustrated in figure 1 and figure 2 requires further investigation into the time dependent in vivo characteristics of this co-polymer.

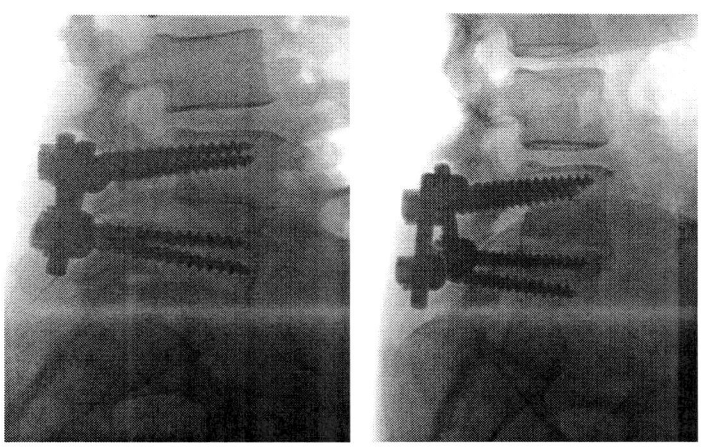

Figure 1: (a) shows direct postoperative radiograph of a 58 year old patient after a PLIF procedure using a 70:30 PLDLLA cage implant impacted with autologous bone. (b) Radiograph of the same patient at 6 months shows obvious signs of subsidence with loss of disc height.

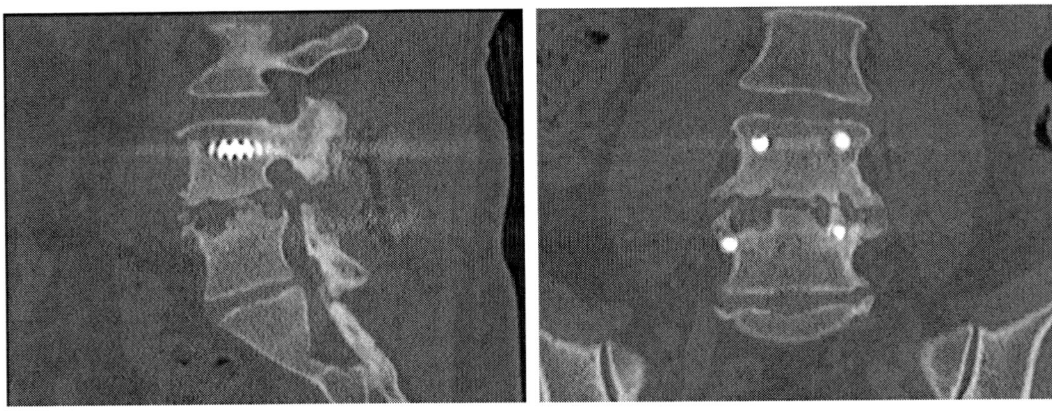

Figure 2 shows CT-scans of the same patient illustrated in figure 1 with signs of osteolysis and non-union

It is currently believed that polyhydroxyacids degrade by a non-specific hydrolytic scission of ester bonds.(Grizzi et al., 1995; Leenslag et al., 1987; Li & McCarthy, 1999; Li, 1999; Loo et al., 2006; Williams, 1992; Ali et al., 1993) This makes them sensitive to moisture and temperature during processing, sterilization and packaging prior to implantation. PLA degrades to form lactic acid which in turn forms pyruvate by lactate dehydrogenase whereas PGA degrades to glycine. Both pyruvate and glycine can enter the tricarboxylic acid cycle and be excreted in the form of water and CO_2. Degradation products of PLA and PGA do not exhibit any significant immunogenic or carcinogenic effects. However, since the degradation products are acidic, it has been argued that the local pH of the surrounding tissue may be elevated as a consequence.(Bostman & Pihlajamaki, 2000a; Bostman & Pihlajamaki, 2000b) Also, macrophages infiltrating in response to foreign degradation material are known to release free radicals and degenerative enzymes which affect both implant and surrounding tissues, further lowering the local pH. Osteoclasts are known to become more active in lower pH and this may explain the mechanism of adverse tissue reactions such as osteolysis sometimes seen with these implants. Contrary to the proffered argument, van der Elst et al. reported no pH alterations during degradation of large PLA and PLA/PGA implants within the intramedullary canal of the sheep femur.(van der Elst et al., 1999)

It remains of interest as to whether the nature and rate of degradation of resorbable implants is site specific or in other words, dependent on the anatomical and physiological characteristics of the implantation site.(Bhatia et al., 1994) The rate of degradation is extremely important particularly with respect to spinal implants most of which are intended to bear significant axial loads, bending and shear forces. The higher rate of degradation seen in implants fabricated from PGA as compared to PLA based implants, may explain the rapid accumulation and high levels of degradation products, hence accelerating the mechanisms that lead to adverse tissue reactions and osteolysis.(Cutright & Hunsuck, 1971; Lippman et al., 2004) Degradation rate is affected not only by the average molecular weight and crystallinity, but also by the geometry of the implant.(Grizzi et al., 1995) A thick walled implant will trap acidic products within the central amorphous structure which may induce autocatalysis, whereas a thin walled implant will permit a more graduated diffusion of these products to the periphery, enhancing its elimination within the vascular matrix. Kandziora et al. reported rapid disintegration, collapse and adverse tissue reaction only after 12 weeks of implantation of a thick walled (±5mm) PLDLLA cervical cage in sheep,(Kandziora et al.,

2004) a phenomenon not seen in a long-term (24 months) in vivo animal study by Toth et al. using thin walled (±2mm) PLDLLA cages.(Toth et al., 2002)

With the current knowledge available on biocompatibility and time engineered resorption characteristics, human application of PLA and PGA based spinal implants can be considered safe. There is however reason to remain cautious and alert based on earlier reports on possible local adverse tissue reaction associated with these implants.

BIOMECHANICAL ASPECTS

The biomechanical properties of both PLA and PGA have been studied extensively.(Ames et al., 2002a; Ames et al., 2002b; Ames et al., 2005a; Ames et al., 2005b; Ames et al., 2005c; Deguchi et al., 1998; DiAngelo et al., 2002a; Kandziora et al., 2002; Khodadadyan-Klostermann et al., 2001; Pflugmacher et al., 2004; Shikinami & Okuno, 2003; Totoribe et al., 2003; van Dijk et al., 2002c; van Dijk et al., 2003) PLLA and PLDLLA (70/30 poly-L-lactide-Co-D,L-lactide) co-polymer have received the most attention and appear to possess the greatest potential for utility as implants intended for structural support application in spinal surgery. An ideal implant must withstand the biomechanical demands of the human spine until fusion has occurred, whilst exhibiting a modulus of elasticity close to that of bone, and enhance biological spinal segment fusion comparable to existing alternatives. Both PLLA and PLDLLA co-polymer can be engineered to possess adequate strength and stiffness whilst permitting gradual load transfer to healing graft bone.

In vitro and in vivo biomechanical testing has been performed mainly with interbody cages (spacers), tension band anterior plates and posterior (pedicle, laminar and facet) screws and connecting rods. Several studies have demonstrated the potential advantage of the proximity of the modulus of elasticity of resorbable implants to that of bone making these implants more suitable for fusion procedures when compared to metallic counterparts.(van Dijk et al., 2002c; van Dijk et al., 2003; Smit et al., 2003; Wuisman et al., 2002) Human vertebral bone tissue has a modulus of elasticity of 2GPa whereas stainless steel and titanium alloy implants have a modulus of elasticity of 200GPa and 110GPa respectively. PLLA on the other hand with a modulus of elasticity of 4.2GPa better approaches the biomechanical characteristics of vertebral bone tissue making it suitable for utility as structural support device. Adequate loading of the graft is achieved through minimal stress shielding and subsidence is limited.

In a study published by the skeletal tissue engineering group Amsterdam (STEGA), PLLA interbody fusion cages were fabricated and tested using the yield strength of native goat motion segments as reference parameter. Compression testing demonstrated that the designed PLLA cages possessed adequate mechanical characteristics soon after implantation, with or without bone graft containment.(van Dijk et al., 2003) Insertion of the PLLA cage did not negatively influence the compressive strength of the motion segment neither did standardized perforation of the vertebral end plate affect the PLLA constructs' compression strength. An in vivo investigation by van Dijk et al. demonstrated that the lower stiffness of PLLA cages significantly enhanced lumbar interbody fusion when compared to traditional titanium cages.(van Dijk et al., 2002c) They also reported that none of the PLLA cages lost its mechanical integrity 6 months after implantation.

70:30 PLDLLA, when optimally engineered, has also been shown to possess adequate initial mechanical properties for utility as a spinal structural support construct.(Krijnen et al., 2006; Smit et al., 2006b) Whereas PLLA cages have been shown to withstand the biomechanical demands of the spine as stand alone construct, data on PLDLLA calls for caution in this respect.(Krijnen et al., 2006; Smit et al., 2006a; Smit et al., 2006b) The cages have been observed to show fractures after only three months of implantation in a goat, although mechanical strength was shown to be maintained for more than six months in vitro.(Krijnen et al., 2004; Krijnen et al., 2006; Smit et al., 2006a; Smit et al., 2006b) After six months, the PLDLLA cages were strongly deformed and fractured and in half of the goats, non-fusion had developed with severe fibrous tissue formation within the fusion zone. The operated segment apparently was left unstable after mechanical failure of the cage. The reason for premature mechanical failure of the cages appears to be the time-dependent behavior typical for polymers: strength strongly depends on temperature and loading rate, and continuous loading under forces well below the ultimate strength also can lead to failure.(Sturm et al., 2006) Loading at some 50% of the ultimate strength, cage failure occurred after less than two hours. Clearly, time-dependent failure of resorbable polymers under mechanical loading is an issue that needs to be addressed for polymer applications under higher mechanical loading conditions.

Based on current research data, the use of anterior or posterior instrumented augmentation with PLDLLA cages appears to be mandatory. Pflugmacher et al. tested 40 intact cervical sheep spines in flexion, extension, axial rotation and lateral bending with a non destructive stiffness method. Subsequently a PLDLLA cage was implanted in 10 specimens followed by the same biomechanical evaluation.(Pflugmacher et al., 2004) Finally all 10 specimens received additional plating after which renewed biomechanical testing were conducted. In comparison to intact motion segment, the PLDLLA cage showed a significantly higher range of motion and a significantly lower stiffness in rotation. Additional anterior plate stabilization significantly decreased the range of motion and increased the stiffness. There was no biomechanical difference between the PLDLLA cage and the intact motion segment in flexion, extension and bending. Interbody cervical fusion cages fabricated from composites of PLLA and hydroxyapatite have been mechanically tested and found to possess compressive strength that surpasses that of existing metal and carbon fiber cages.

Current pre-clinical and clinical data suggest that resorbable cage implants possess adequate biomechanical properties and may in fact offer biomechanical advantages due to the lower stiffness and closeness of the modulus of elasticity to that of bone. The notion with respect to resorbable screws, pins and rods seems however to be less promising. Deguchi et al. compared the biomechanical performance of trans-laminar facet joint fixation using either 2mm PLLA pins or 3,5mm stainless steel cortical screws in 9 cadeveric sheep spine (L2-L6).(Deguchi et al., 1998) Each spinal segment was non-destructively tested in flexion-extension bending before and after screw or PLLA pin fixation. They concluded that the use of PLLA trans-laminar pins offers some stability to the lumbar spine but is significantly less stiff when compared to screw fixation. In a paper published by Ella et al. in 2005, self reinforced PLLA screws and a novel resorbable intervertebral disc implant with an elastic core surrounded by matrix and reinforcement material were tested for mechanical properties, molecular weight changes and thermal properties in vitro.(Ella et al., 2005) The elastic nucleus of the cylindrical disc implant was made of 96:4 poly(L/D)lactide, 70:30 poly(L/DL)lactide, bioactive glass and polyactive 1000PEOT70PBT30 and all implants were gamma sterilized. Both implants showed no changes in the compression properties throughout in vitro testing. Based on the adequate mechanical properties demonstrated, they concluded that both implant types could be used in clinical testing.

A novel resorbable polylactide co-polymer odontoid screw has been evaluated for its ability to restore strength and stiffness to the fractured odontoid process compared with traditional titanium screws.(Ames et al., 2005c) The resorbable screw was fabricated from 70:30 PLDLA, cannulated and 4.5mm in diameter. A fracture of the odontoid was induced in 14 cadeveric human spinal specimens and subsequently one half of the fractured odontoid was fixed with a resorbable screw and the other half with a titanium screw. The ultimate strength defined as the maximum force sustained before failure was similar with both screws, and significantly lower than the force to failure of the intact odontoid prior to fracture. The stiffness of the fractured odontoid process was restored to 15 and 23% of its original value by repairing it with resorbable and metal screws, respectively. Handolin et al. investigated the effects of low intensity pulsed ultrasound on 96 resorbable self reinforced PLLA screws by measuring the bending strength, shear strength and molecular weight after exposure to ultrasound.(Handolin et al., 2002; Handolin et al., 2005) There were differences in the investigated properties between ultrasound exposed screws and non-exposed control screws. They concluded that resorbable PLLA screws are compatible with low intensity pulsed ultrasound.

In general, biomechanical data on resorbable spinal implants reveals promising results with sufficient initial strength and an elastic modulus approaching that of bone tissue. These characteristics are often retained by the resorbable implant long enough to allow fusion, whilst subsequent resorption permits gradual transfer of load to the fusion mass promoting biological remodeling. However, polylactide strength appears to depend strongly on loading rate, and time-to-failure is considerably reduced under continuous loading conditions. Additional fixation appears to be mandatory for application of the polymer in spinal cages.

Preclinical Experiences

Extensive animal and cadaveric experimental studies have been published highlighting the potential benefits of resorbable spinal implants. Most data concentrate on implants intended to aid spinal fusion of which resorbable interbody cage implants appear to be the most promising and consequently has received the most attention. The skeletal tissue engineering group Amsterdam (STEGA) has published several papers on preclinical research of resorbable lumbar interbody cages.(Krijnen et al., 2004; Krijnen et al., 2006; Smit, 2002; Smit et al., 2003; Smit et al., 2006a; Smit et al., 2006b; Tunc et al., 2004; van Dijk et al., 2002a; van Dijk et al., 2002b; van Dijk et al., 2002c; van Dijk et al., 2002d; van Dijk et al., 2003; van Dijk et al., 2005; Wuisman et al., 2002; Wuisman & Smit, 2006) Based on the results of biomechanical testing on single motion segments of 21 Dutch milk goats, the research team at STEGA designed and fabricated pure poly-L-lactic acid (PLLA) cages to meet the required strength and stiffness as to be able to withstand the compression, shear and bending forces of the spine.(van Dijk et al., 2002a) In 2002 a paper from this group presented the results of fusion of a lumbar segment in dutch milk goats using resorbable PLLA cage implants.(van Dijk et al., 2002c) Two different profiles (stiff and flexible) of PLLA cages were implanted (3 each) and compared to titanium cages. Upon conclusion of the study at 6 months no PLLA cage was found to have lost its mechanical integrity nor did any specimen show radiographic evidence of collapse. Fusion rate was significantly higher in the grouped PLLA implanted goats than those implanted with titanium cages whilst subsidence was significantly lower with titanium cages. PLLA cages often demonstrated solid fusion with bridging trabecular bone whereas titanium cages often showed in growth of trabecular bone with discontinuity in the fusion mass, a pattern seen in the clinical setting and often referred to as locked pseudoarthrosis. In a 3 year in vivo study by the same group(Tunc et al., 2004) the authors found fusion to be significantly faster and more complete when using PLLA cages compared to titanium cages with the same dimensions. There were no signs of subsidence or implant collapse in the PLLA group, and bone remodelling within the cage was completed 2 years after implantation. In terms of degradation of the PLLA, similar features were observed in vivo and in vitro with degradation almost completed 3 years after implantation. Tissue reaction was mild during the 3-year period. Complete resorption of the PLLA cage was observed in half of the specimens, and in the remaining specimens an estimated 1-10% of the original PLLA was present at the 3-year follow-up. In a second paper reporting on the same study extended to a 4 year period(van Dijk et al., 2005) sequential histologic analysis of instrumented motion segments, lymph nodes, and nervous structures showed no adverse local or distant histological or systemic effects during the absorption of the PLLA cages. Interbody fusion was maintained, and only a very mild inflammatory response was observed. At the 4-year follow-up, five out of seven PLLA specimens showed no PLLA particles under polarized light microscopy. In the remaining two specimens an estimated 1% of the original PLLA could be observed. PLLA cages were proven to be feasible for lumbar interbody fusion, with excellent biocompatibility and complete absorption of the PLLA interbody fusion cages can be expected at 3 to 4 years after implantation.

Contrary to PLLA cages, 70:30 poly(L-lactide-co-D,L-lactide) (PLDLLA) cages seem to have a high rate of failure when utilized as a stand alone cage implant in lumbar spinal fusion. On-going animal studies by the research team at STEGA indicate that PLDLLA cages when implanted as stand alone to facilitate lumbar interbody fusion provided insufficient mechanical stability within the first 6 months.(Smit et al., 2006a) The implant tends to exhibit cracking and deformation with significantly lower fusion rates. Supplemental internal fixation improves mechanical integrity and leads to fusion rates comparable to existing alternatives. A mild tissue reaction has been observed in response to degradation products of PLDLLA cages.

The same group have studied the effect of various sterilization techniques on the mechanical integrity and consequently rate of degradation of PLDLLA cage implants. There is sufficient evidence to believe that E-beam sterilization strongly decreases the inherent viscosity of 70:30 PLDLLA cages when compared to EtO sterilized cages, although the mechanical strength seems to be only marginally affected in the initial 6 months.(Loo et al., 2005a; Loo et al., 2005b; Loo et al., 2006) After 6 months the mechanical strength of E-beam sterilized cages becomes increasingly affected and reduces to negligible values by 12 months, whereas EtO sterilized cages retained more than 80% of its initial mechanical integrity.(Smit et al., 2006b) It does appear that depletion of inherent viscosity which in fact reflects diminishing molecular weight must attain a certain threshold in order to lead to rapidly progressive mechanical failure. There is continued need for caution in the use of 70:30 PLDLLA cages as stand alone lumbar spinal implants.

The efficacy of 70:30 PLDLLA cages packed with iliac crest autograft in enhancing lumbar spinal fusion has been compared to that of similar cages packed with resorbable collagen material impregnated with human recombinant bone morphogenic protein. Toth et al. published a paper using the sheep spine model in which interbody fusion at L4-5 level was performed using either PDLLA cage packed with morselized iliac crest graft or infuse bone graft substitute consisting of 0.8ml of rhBMP-2 applied to a type I collagen sponge.(Toth et al., 2002) Rate of fusion was similar in both groups and device degradation was associated with mild to moderate chronic inflammatory response at all postoperative sacrifice times. An ongoing similar study by the STEGA using recombinant human osteogenic protein (rhOP-1) however indicates a tendency toward significantly lower fusion rates in PLLA cages packed with rhOP-1 compared to PLLA cages packed with autologous iliac crest graft.

Several animal experimental studies using resorbable cages to enhance cervical interbody fusion have also been conducted and reported on. Cahill et al. conducted a pilot study and determined that a resorbable cage composed of 85:15 polylactide-polyglycolic co-polymer was not suitable for cervical interbody fusion after anterior discectomy.(Cahill et al., 2003) The same conclusion was reached by Lipmann et al. who observed that the 85:15 polylactide-polyglycolic co-polymer cages resorbed too quickly.(Lippman et al., 2004) Another resorbable cage composed of 70:30 PLDLLA/PGA was however found by the same group to work as well as tri-cortical autograft when filled with cancellous autograft and better still when filled with rhBMP-2.

Data on the application of resorbable plates, rods, screws and pins are scarce. The ability to heat resorbable plates and rods above the glass transition temperature and subsequently contour them to fit the curvature of the spinal segment in question makes them desirable for application as structural support devices. DiAngelo et al. demonstrated improved mechanical stability obtained upon additional anterior plating with a resorbable implant after traditional metal cage implantation through an anterior approach in the cadaveric lumbar spine.(DiAngelo et al., 2002b; DiAngelo et al., 2002a) Plating restores the tension band effect of the anterior longitudinal ligament which is sacrificed at surgery. The increased mechanical integrity is most obvious in extension loading. The resorbable plates used were fabricated from amorphous PLDLLA and E-beam sterilized. Promising results were also reported by Cornwall et al. who used an ovine model in which a single level cervical fusion was performed with tri-cortical iliac crest graft and resorbable plates and screws. The resorbable implants were also made from an amorphous co-polymer of 70:30 PLDLLA. In addition to adequate fusion, biomechanical evaluation demonstrated a significant reduction of motion in all 3 planes at 6 and 12 months after implantation. No non-plated control sheep was used because previous evaluations have led to graft extrusion and neurological compromise in these animals.

A novel resorbable protective graft containment device fabricated from polylactide polymer has been shown to enhance posterolateral fusion using either autologous bone graft or demineralised bone matrix in rabbits. Poynton et al. performed posterolateral lumbar fusion in 20 New Zealand white rabbits using either autograft or demineralised bone matrix with or without a containment device.(Poynton et al., 2002b; Poynton et al., 2002a) The quality of the fusion mass was superior in the specimens with a resorbable graft containment device. Three dimensional CT-scan showed the fusion mass to be cylindrical and dense when a graft containment device was used as opposed to the flat sheet-like fusion mass seen when the containment device was not used.

Lots of experience has been gathered with the use of resorbable rods and screws in fractures of the extremity but its application in spinal surgery is very limited with only a handful preclinical study reported and no clinical studies to date. Bezer et al. recently published a paper in which a self reinforced polylactide (SR-PLLA) rod was used to stabilize the lumbar spine in a posterolateral fusion procedure in a rabbit model.(Bezer et al., 2005) There was 100% fusion in all 8 rabbits at 12 weeks evaluation whereas only one fusion was seen in the non-instrumented control group. In this study, the use of SR-PLLA rods proved to be as effective as stainless steel rods.

The suitability and application of resorbable polymer sheets as protective and containment devices for either bone grafts or neural tissues have been reported. In a study by Welch et al. two thicknesses of resorbable PLA film was implanted as an adhesion barrier in an ovine unilateral laminotomy model.(Welch et al., 2002) The readily contourable thinner film was placed directly over the dura whereas the thicker film was placed above a lamina defect to seal off the epidural space. The resorbable films were shown to prevent bony re-growth at the vicinity of the laminotomy window, prevent soft tissue prolapse through the laminotomy window and prevent tenacious adherence of the dura to the surrounding tissues. Further analysis showed considerable reduction in total collagen within the scar tissue samples as compared to the untreated control samples. More recently, an ovine laminectomy model was used to evaluate the efficacy of a polylactide barrier film in a larger defect by performing a dorsal laminectomy and durotomy.(Klopp et al., 2004) The material tested as a barrier was a 0.02mm thin resorbable film fabricated from 70:30 poly(L-lactide-co-D,L-lactide). No compromised healing or cerebrospinal fluid leakage was found on myelography, a potential complication previously reported with an existing alternative ADCON-L. Both the tenacity and volume of the peridural adhesions were considerably decreased and histological studies revealed no signs of adverse inflammatory response.

Oudega et a. investigated axonal growth and myelination in a schwann cell graft contained in polylactide tubular scaffolds up to 4 months after implantation in the completely transected adult rat thoracic spinal cord.(Oudega et al., 2001) They found that axonal growth was promoted within the scaffolds with unmyelinated axons and blood vessels found in the schwann cell grafts as early as 2 weeks after implantation. Myelination was demonstrated at one month and beyond. Of the 2 different tubular scaffolds used, the tube fabricated from slower degrading mixture of poly(L-lactic acid) and 10% poly(L-lactic acid) oligomers produced a larger regeneration response than the tube fabricated from poly(D,L-lactic acid).

The volume and diversity of preclinical data on the spinal application of resorbable implants has immensely contributed to our understanding of the physical, biological and biomechanical behaviour of these polymers both in vitro and in vivo. Although generally safe and efficacious in the promotion of spinal fusion, and as containment and protective spinal devices, there appears to be a specific risk of potential adverse effects related to the time dependent degradation properties of the various co-polymer ratios used in the fabrication of these implants. Furthermore, their efficacy in regeneration of the spinal cord and other neural elements remains to be proven.

Clinical Experiences

There is currently limited experience with the clinical application of resorbable spinal implants. Vast majority of data concerns implants intended to aid spinal fusion, of which interbody cages have been investigated the most. Two commonly used resorbable co-polymers have been approved by the Food and Drug Administration for orthopaedic and neurosurgical applications. These are the co-polymer poly(L-lactide-co-D,L-lactide) and a second co-polymer PLLA/PGA. Figure 3 shows successful fusion acquired following a PLIF procedure using an interbody cage implant fabricated from poly(L-lactide-co-D,L-lactide).

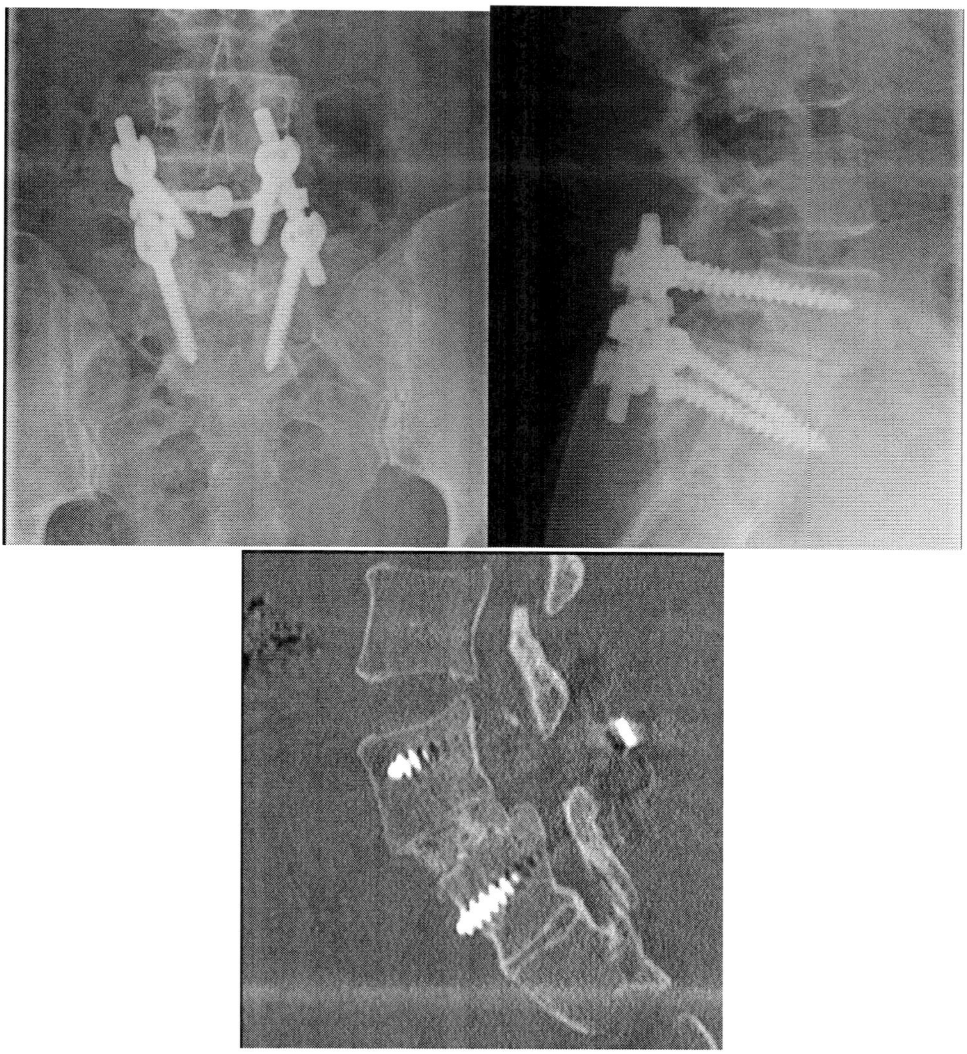

Figure 3 shows solid fusion in a 42 year old patient at 1 year after a PLIF procedure using a 70:30 PLDLLA cage implant.

The first clinical application of resorbable spinal interbody fusion cage was published by Lowe and Coe in 2002.(Lowe & Coe, 2002b; Lowe & Coe, 2002a) The cage was fabricated from a 70:30 ratio of poly(L-lactide-co-D,L-lactide) copolymer and implanted in 60 patients after being impacted with autologous iliac crest graft. This implant retains 100% of its compressive strength 6-9 months after implantation, 97% at 12 months and completely resorbs between 18-36 months. Posterior pedicle screw instrumentation was used in all patients. The follow up was however short, one to nine months (mean 4.7) and although one to five spinal segments per patient were fused, cages were not inserted into all levels. They reported symptomatic relief in a majority of the patients, with minimal postoperative complications and no implant failure. There was however one case of cracked implant upon insertion although no subsidence or other evidence of structural failure was observed in this patient. Specifically, there have been no adverse events either directly or indirectly related to the use of the resorbable implant. Subsequent report by the same authors in 2004 and 2005 confirmed their preliminary report at a minimum follow up of 2 years, with a fusion rate of 92.6% whilst 81.5% of patients had a good or excellent result based on Prolo's criteria.(Coe, 2004; Coe & Vaccaro, 2005) It is of interest to note that this long term follow up showed no adverse effects either directly or indirectly related to the resorbable implants used.

Despite very promising results from this group of authors, we recommend cautious interpretation for the following reasons. The methodology is inadequate taking into account the fact that a relatively new product is being investigated. It is not clear from all three papers if the set up of the study was prospective, if the patients were consecutively enrolled, and the patients were not randomized to compare with existing treatment options. The surgical indication was too diverse to draw any conclusions for one particular indication, and the number of spinal segments fused was not uniform. Assessment of fusion was based only on radiographs although the superiority of thin slice computer tomographs have been reported, an observation being made by the STEGA in an ongoing prospective randomized single blinded study comparing PLIF with either a resorbable interbody cage or a non resorbable implant. Furthermore the assessment of patient centered clinical outcome measures is not representative since only 16 of the initially reported 60 patients partook in the SF-36 outcome analysis. Consequently, these three pioneer studies do not help determine which patient group (surgical indication) might benefit from this new implant. It does not offer reliable values of either fusion rates nor patient centered clinical outcome and does not offer an adequate basis for comparison with future studies.

In 2003 Austin et al. published a paper based on a series of 12 patients with diverse spinal pathologies treated by PLIF using Hydrosorb interbody cages impacted with autograft.(Austin et al., 2003) The cage used was a poly(L-lactide-co-D,L-lactide) copolymer and similar to that used in a previous study by Lowe and Coe. All patients received multiple segment fusion (range 2-4) and had posterior pedicle screw instrumentation in addition. Although all patients had satisfactory clinical results and went on to complete fusion, the number of patients is too small to make any definitive recommendations. Also assessment of fusion was not consistently defined for all patients and no patient centered clinical outcome measure was utilized. Again interestingly no implant related adverse event was reported.

Couture and Branch from the same institution published fusion rates and clinical outcome of a series of 27 patients who underwent PLIF with instrumented pedicle screw fixation and hydrosorb interbody cages.(Couture & Branch, Jr., 2004) A rectangular shaped cage was used in this study as opposed to the cylindrical cage used in the studies published by Lowe and

Coe. The indication for surgery was uniform and consistent; all patients presented with a combination of mechanical low back pain, neurogenic claudication and significant radicular component refractory to non-surgical measures. A 360 degrees fusion was performed in all patients. They reported 100% fusion rate in 19 patients who received one or two level fusion, 93.3% fusion rate in 5 patients with three levels fused and 75% fusion rate in one patient with 4 levels fused. It is however important to note that those patients with an incomplete fusion on CT-scan but no clinical or radiological signs of pseudoarthrosis were assigned to the successful fusion group. It is therefore possible that the true rate of fusion has been overstated. Although failure to fusion was manifested by clinical deterioration and varying degrees of disc collapse, cavitations and lucency around the Hydrosorb cages, no evidence of an adverse inflammatory response was found at revision surgery. It is however unclear if histology was performed to confirm the presence and extent of inflammatory response and degradation particles of the cage.

Clinical studies using resorbable polymers as containers for osteogenic growth factors in achieving spinal fusion has also been reported. Lanman and Hopkins published two papers in 2004 one of which included 43 patients who had a TLIF procedure of the lumbar spine using Hydrosorb cage containing collagen sponge impregnated with recombinant human bone morphogenic protein-2.(Lanman & Hopkins, 2004b; Lanman & Hopkins, 2004a) The second paper included 20 patients who underwent anterior cervical discectomy and interbody fusion (ACDF) using Conerstone HSR resorbable implant, a non-crystalline 70 30 poly(L-lactide-co-D,L-lactide) containing collagen sponge impregnated with recombinant human bone morphogenic protein-2. They reported 98% fusion rate at 6 months follow up in the lumbar spinal interbody fusion group and 100% fusion rate at 3 months in the ACDF group. Patient centered clinical outcome measures including oswestry disability and SF-36 scores showed an overall improvement in both studies. There were no adverse implant related effects reported.

One of the most promising clinical results up to date is perhaps that published by Kulko et al. in which a consecutive series of 22 patients underwent TLIF procedure using resorbable Hydrosorb interbody cages packed with recombinant bone morphogenic protein.(Kuklo et al., 2004) Radiographic fusion rate was 87.2% at a mean follow up of 12.4 months. This was confirmed on 3D CT scan evaluation in which a 97.3% fusion rate was reported. Unfortunately no patient centered clinical outcome measure was utilized.

Virtually all clinical studies published till date seem to support the notion that spinal application of resorbable polymers is not associated with adverse local tissue reactions, yet recent non-published reports by colleagues in Europe and America echoes a warning on the potential risks of massive osteolysis, subsidence and collapse of these implants in vivo. A paper presented by Herceg at the 19th annual meeting of the North American Spine Society in Chicago in 2004 highlighted the problem of rapid subsidence in lumbar interbody fusions seen when resorbable polymer implants are applied.(Herceg et al., 2004) These reports call for caution with widespread clinical application of these implants outside the scope of controlled trials.

The potential benefits of high resolution MRI in the evaluation of fusion following interbody resorbable cage implantation have been demonstrated in recent animal studies.(Krijnen et al., 2004) PLA polymer based resorbable implants do not degrade the quality of MRI therefore, MRI can be used to evaluate the tissue response to the implant as well as assess implant integrity and degradation at various time intervals. PLA implants are visible as areas homogenous low signal intensity, which can be distinguished from the high signal intensity of surrounding bone. Upon subsequent degradation plastic deformation and cracks can be noted on transverse MR images and as hydrolysis progresses, the implant becomes impregnated with water leading to high signal intensity on MR imaging.

Johnsson et al. used roentgen stereophotogrammetric analysis (RSA) to study the kinematics and healing rate of non-instrumented posterolateral lumbar spinal fusion with additional trans-articular facet joint fixation using resorbable self reinforced polyglycolic acid rods(Johnsson et al., 1997). All 7 patients with a single level procedure went on to complete fusion. Three of the remaining 4 patients with multi level procedure showed no signs of complete fusion at one year follow up. One patient was excluded due to improper placement of the tantalums balls. None of the 11 cases showed any signs of osteolysis within the region of the implanted resorbable rod.

In summary, extensive research has been done and is still ongoing in the use of resorbable implants in spinal surgery. Currently available clinical data show promising results with fusion rates comparable to historical data on existing non-resorbable implants and more importantly, there appears to be no risk of direct or indirect implant related adverse effects. Data reflecting patient improvement based on patient centered outcome measures is however lacking. Furthermore all studies up till date offer at best level IV evidence with respect to the efficacy of resorbable implants intended for achieving spinal fusion, and even less clinical data is available regarding the utilization of resorbable screws, pins, rods and neural protective devices in spinal surgery.

The current literature is therefore far from adequate and inconclusive. There is need for proper guidelines as to which patient population will benefit the most with the use of these new implants. More knowledge needs to be gained on the thoroughness of preparation of the implantation site (endplates) and the time related effects of the static and dynamic loading on the time dependent biomechanical properties of the implant material. Sporadic reports on significant subsidence (figure 1) and massive osteolysis (figure 2) seen in some patients that have undergone PLIF with the 70:30 poly(L-lactide-co-D,L-lactide) cage is worrisome and needs investigation.

Future Directions in Spine Related Resorbable Technology

Although major advances have been booked on the engineering and fabrication of resorbable implants to meet the various biological and biomechanical demands of the human spine, there are still several questions that need to be answered. The optimal polymer and co-polymer composition characteristics have yet to be defined for the various spinal application devices. Several mixtures of PLA and PGA co-polymers, ranging from pure PLLA to 70:30 PLDLLA and 85:15 PLDLLA/PGA have been used to fabricate varying spinal protective and structural support devices without a clear direction as to which mixture best serves the need of a specific implant. The mode of failure of PLA based implants seems to be time-dependent and influenced by the nature of static and dynamic loading in vivo. Current design parameters are obtained mainly from in vitro biomechanical testing which has an inherent drawback of not being able to entirely mimic the simultaneous and complex nature of static and dynamic loading in vivo. Consequently future research should be directed towards models that will help understand and predict not only the biological behavior but also the biomechanical behavior and performance of these polymers in vivo.

Furthermore, there is need to clarify the influence of anatomical site of implantation on the behavior of the implant with a goal towards the development of site specific implants. Previous animal studies have indicated that similar devices when implanted into different tissues or anatomical locations, may exhibit different biological and biomechanical behavior. Upon achieving the goal of a site specific implant with accurately predictable biological and biomechanical performance, one can now move unto the conduction of appropriate clinical research to establish the safety and efficacy of the implant.

The quality of clinical research work published thus far is poor and little evidence has been provided for the efficacy of resorbable spinal implants in structural support applications and protective neural devices. Although clinical work has only recently been published and still few in number, virtually no study has well defined methodology, indications for surgery is too diverse and the number of spinal segments fused is not uniform. No clinical study has prospectively randomized patients and compared the relatively new resorbable spinal implants to existing alternatives be it non-resorbable implants or non-operative treatment options for specified pathologic spinal conditions. Current clinical studies have failed to demonstrate the proffered advantage of resorbable spinal implants with respect to imaging quality and assessment of fusion. Resorbable spinal implants do not degrade the quality of MRI and CT scan and are visible as areas homogenous low signal intensity, which can be distinguished from the high signal intensity of surrounding bone. However, metallic implants including that made from titanium are known to interfere with and degrade the quality of MR imaging, making it difficult to perform diagnostic neural imaging in the region of a spinal arthrodesis. Even when no implants are used, excessive scar tissue can hamper diagnostic MRI, whereas resorbable protective barrier sheaths have been shown to significantly reduce perineural scarring. Furthermore it has been advocated that thin sliced CT scanning is more effective in establishing complete spinal fusion compared to conventional radiographs.

Unfortunately the majority of clinical studies on resorbable spinal implants thus far have failed to utilize CT scans and MRI to evaluate patients postoperatively and report on these proffered benefits. The authors are currently conducting a prospective randomized single blinded study comparing the efficacy of PLIF with posterior instrumentation in which either a resorbable or a non resorbable cage is implanted in a single spinal segment, in patients with symptomatic spondylolysis and spondylolisthesis, with or without neurologic compromise. All patients are assessed using patient centered clinical evaluation as well as imaging modalities including radiographs, CT scans and MRI and all patients will undertake a minimum of 2 year follow up.

CONCLUSION

The application of surgical implants in the treatment of the various spinal ailments is still on the rise. When non resorbable surgical implants are used there is a potential risk not only of early implant failure but also of long-term complications attributable to the implant. Furthermore, retrieval studies have shown that when metal and carbon fibre reinforced implants fail, the local effects on bone and surrounding neural tissues could be deleterious.

The concept of applying resorbable implant technology to achieve the same goal and thus eliminating the potential long term implant related complications is therefore a desirable one. This concept has been made feasible by the application of resorbable polymer technology to fabricate implants that possess adequate physical and biological properties to withstand the demands of the human spine. Despite major progress made so far, several worrying issues remain to be resolved. The time-dependent effects of static and dynamic loading of these implants on their degradation in vivo needs further investigation, as this may play a role in the failure mode resulting in subsidence and osteolysis sometimes associated with these implants.

REFERENCES

Alexander, J. T., Branch, C. L., Jr., Subach, B. R., and Haid, R. W., Jr. (2002). Applications of a resorbable interbody spacer in posterior lumbar interbody fusion. *J Neurosurg. 97*, 468-472.

Ali, S. A., Doherty, P. J., and Williams, D. F. (1993). Mechanisms of polymer degradation in implantable devices. 2. Poly(DL-lactic acid). *J Biomed Mater Res,* 27, 1409-1418.

Ames, C. P., Acosta, F. L., Jr., Chamberlain, R. H., Larios, A. E., and Crawford, N. R. (2005a). Biomechanical analysis of a newly designed bioabsorbable anterior cervical plate. Invited submission from the joint section meeting on disorders of the spine and peripheral nerves, March 2005. *J Neurosurg Spine*, 3, 465-470.

Ames, C. P., Cornwall, G. B., Crawford, N. R., Nottmeier, E., Chamberlain, R. H., and Sonntag, V. K. (2002a). Feasibility of a resorbable anterior cervical graft containment plate. *J Neurosurg*, 97, 440-446.

Ames, C. P., Crawford, N. R., Chamberlain, R. H., Cornwall, G. B., Nottmeier, E., and Sonntag, V. K. (2002b). Feasibility of a resorbable anterior cervical graft containment plate. *Orthopedics*, 25, s1149-s1155.

Ames, C. P., Crawford, N. R., Chamberlain, R. H., Deshmukh, V., Sadikovic, B., and Sonntag, V. K. (2005b). Biomechanical analysis of a resorbable anterior cervical graft containment plate. *Spine*, 30, 1031-1038.

Ames, C. P., Crawford, N. R., Chamberlain, R. H., Deshmukh, V., Sadikovic, B., and Sonntag, V. K. (2005c). Biomechanical evaluation of a bioresorbable odontoid screw. *J Neurosurg Spine*, 2, 182-187.

An, Y. H., Woolf, S. K., and Friedman, R. J. (2000). Pre-clinical in vivo evaluation of orthopaedic bioabsorbable devices. *Biomaterials*, 21, 2635-2652.

Austin, R. C., Branch, C. L., Jr., and Alexander, J. T. (2003). Novel bioabsorbable interbody fusion spacer-assisted fusion for correction of spinal deformity. *Neurosurg Focus*, 14, e11.

Bendix, D. (1998). Chemical synthesis of polylactide and its copolymers for medical applications. *Polymer Degradation and Stability*, 59, 129-135.

Bergsma, J. E., de Bruijn, W. C., Rozema, F. R., Bos, R. R., and Boering, G. (1995a). Late degradation tissue response to poly(L-lactide) bone plates and screws. *Biomaterials*, 16, 25-31.

Bergsma, J. E., Rozema, F. R., Bos, R. R., Boering, G., de Bruijn, W. C., and Pennings, A. J. (1995b). Biocompatibility study of as-polymerized poly(L-lactide) in rats using a cage implant system. *J Biomed Mater Res*, 29, 173-179.

Bezer, M., Yildirim, Y., Erol, B., and Guven, O. (2005). Absorbable self-reinforced polylactide (SR-PLLA) rods vs rigid rods (K-wire) in spinal fusion: an experimental study in rabbits. *Eur Spine J*, 14, 227-233.

Bhatia, S., Shalaby, S. W., Powers, D. L., Lancaster, R. L., and Ferguson, R. L. (1994). The effect of site of implantation and animal age on properties of polydioxanone pins. *J Biomater Sci Polym Ed*, 6, 435-446.

Blumenthal, S. L. and Ohnmeiss, D. D. (2003). Intervertebral cages for degenerative spinal diseases. *Spine J*, 3, 301-309.

Bostman, O. and Pihlajamaki, H. (2000a). Clinical biocompatibility of biodegradable orthopaedic implants for internal fixation: a review. *Biomaterials*, 21, 2615-2621.

Bostman, O. M., Laitinen, O. M., Tynninen, O., Salminen, S. T., and Pihlajamaki, H. K. (2005). Tissue restoration after resorption of polyglycolide and poly-laevo-lactic acid screws. *J Bone Joint Surg Br*, 87, 1575-1580.

Bostman, O. M. and Pihlajamaki, H. K. (1998). Late foreign-body reaction to an intraosseous bioabsorbable polylactic acid screw. A case report. *J Bone Joint Surg Am*, 80, 1791-1794.

Bostman, O. M. and Pihlajamaki, H. K. (2000b). Adverse tissue reactions to bioabsorbable fixation devices. *Clin Orthop Relat Res*, 216-227.

Cahill, D. W., Martin, G. J., Jr., Hajjar, M. V., Sonstein, W., Graham, L. B., and Engelman, R. W. (2003). Suitability of bioresorbable cages for anterior cervical fusion. *J Neurosurg*, 98, 195-201.

Cizek, G. R. and Boyd, L. M. (2000). Imaging pitfalls of interbody spinal implants. *Spine*, 25, 2633-2636.

Coe, J. D. (2004). Instrumented transforaminal lumbar interbody fusion with bioabsorbable polymer implants and iliac crest autograft. *Neurosurg Focus*, 16, E11.

Coe, J. D. and Vaccaro, A. R. (2005). Instrumented transforaminal lumbar interbody fusion with bioresorbable polymer implants and iliac crest autograft. *Spine*, 30, S76-S83.

Cook, S. D., Patron, L. P., Christakis, P. M., Bailey, K. J., Banta, C., and Glazer, P. A. (2004). Comparison of methods for determining the presence and extent of anterior lumbar interbody fusion. *Spine*, 29, 1118-1123.

Cornwall, G. B., Ames, C. P., Crawford, N. R., Chamberlain, R. H., Rubino, A. M., Seim, H. B., and Turner, A. S. (2004). In vivo evaluation of bioresorbable polylactide implants for cervical graft containment in an ovine spinal fusion model. *Neurosurg Focus*, 16, E5.

Couture, D. E. and Branch, C. L., Jr. (2004). Posterior lumbar interbody fusion with bioabsorbable spacers and local autograft in a series of 27 patients. *Neurosurg Focus*, 16, E8.

Cutright, D. E. and Hunsuck, E. E. (1971). Tissue reaction to the biodegradable polylactic acid suture. *Oral Surg Oral Med Oral Pathol*, 31, 134-139.

de Medinaceli, L., al Khoury, R., and Merle, M. (1995). Large amounts of polylactic acid in contact with divided nerve sheaths have no adverse effects on regeneration. *J Reconstr Microsurg*, 11, 43-49.

Deguchi, M., Cheng, B. C., Sato, K., Matsuyama, Y., and Zdeblick, T. A. (1998). Biomechanical evaluation of translaminar facet joint fixation. A comparative study of poly-L-lactide pins, screws, and pedicle fixation. *Spine*, 23, 1307-1312.

DiAngelo, D. J., Kitchel, S., McVay, B. J., Scifert, J. L., and Cornwall, G. B. (2002a). Bioabsorbable anterior lumbar plate fixation in conjunction with anterior interbody fusion cages. *Orthopedics*, 25, s1157-s1165.

DiAngelo, D. J., Scifert, J. L., Kitchel, S., Cornwall, G. B., and McVay, B. J. (2002b). Bioabsorbable anterior lumbar plate fixation in conjunction with cage-assisted anterior interbody fusion. *J Neurosurg*, 97, 447-455.

Elias, W. J., Simmons, N. E., Kaptain, G. J., Chadduck, J. B., and Whitehill, R. (2000). Complications of posterior lumbar interbody fusion when using a titanium threaded cage device. *J Neurosurg,* 93, 45-52.

Ella, V., Kellomaki, M., and Tormala, P. (2005). In vitro properties of PLLA screws and novel bioabsorbable implant with elastic nucleus to replace intervertebral disc. *J Mater Sci Mater Med*, 16, 655-662.

Gautier, S. E., Oudega, M., Fragoso, M., Chapon, P., Plant, G. W., Bunge, M. B., and Parel, J. M. (1998). Poly(alpha-hydroxyacids) for application in the spinal cord: resorbability and biocompatibility with adult rat Schwann cells and spinal cord. *J Biomed Mater Res*, 42, 642-654.

Gogolewski, S. (2000). Bioresorbable polymers in trauma and bone surgery. *Injury*, 31 Suppl 4, 28-32.

Gogolewski, S., Jovanovic, M., Perren, S. M., Dillon, J. G., and Hughes, M. K. (1993). Tissue response and in vivo degradation of selected polyhydroxyacids: polylactides (PLA), poly(3-hydroxybutyrate) (PHB), and poly(3-hydroxybutyrate-co-3-hydroxyvalerate) (PHB/VA). *J Biomed Mater Res*, 27, 1135-1148.

Gogolewski, S. and Mainil-Varlet, P. (1996). The effect of thermal treatment on sterility, molecular and mechanical properties of various polylactides. I. Poly(L-lactide). *Biomaterials*, 17, 523-528.

Gogolewski, S. and Mainil-Varlet, P. (1997). Effect of thermal treatment on sterility, molecular and mechanical properties of various polylactides. 2. Poly(L/D-lactide) and poly(L/DL-lactide). *Biomaterials*, 18, 251-255.

Gogolewski, S., Mainil-Varlet, P., and Dillon, J. G. (1996). Sterility, mechanical properties, and molecular stability of polylactide internal-fixation devices treated with low-temperature plasmas. *J Biomed Mater Res*, 32, 227-235.

Grizzi, I., Garreau, H., Li, S., and Vert, M. (1995). Hydrolytic degradation of devices based on poly(DL-lactic acid) size-dependence. *Biomaterials*, 16, 305-311.

Handolin, L., Kiljunen, V., Arnala, I., Pajarinen, J., Partio, E. K., and Rokkanen, P. (2005). The effect of low intensity ultrasound and bioabsorbable self-reinforced poly-L-lactide screw fixation on bone in lateral malleolar fractures. *Arch Orthop Trauma Surg*, 125, 317-321.

Handolin, L., Pohjonen, T., Partio, E. K., Arnala, I., Tormala, P., and Rokkanen, P. (2002). The effects of low-intensity pulsed ultrasound on bioabsorbable self-reinforced poly L-lactide screws. *Biomaterials*, 23, 2733-2736.

Hasegawa, S., Ishii, S., Tamura, J., Furukawa, T., Neo, M., Matsusue, Y., Shikinami, Y., Okuno, M., and Nakamura, T. (2006). A 5-7 year in vivo study of high-strength hydroxyapatite/poly(L-lactide) composite rods for the internal fixation of bone fractures. *Biomaterials,* 27, 1327-1332.

Herceg, M., Slosar, P.J., Josey R., et al. (2004). Rapid subsidence in lumbar interbody fusions with resorbable implants. Presented at the 19th annual meeting of the North American Spine Society. Chicago, IL. October 28, 2004.

Hojo, Y., Kotani, Y., Ito, M., Abumi, K., Kadosawa, T., Shikinami, Y., and Minami, A. A. (2005). biomechanical and histological evaluation of a bioresorbable lumbar interbody fusion cage. *Biomaterials*, 26, 2643-2651.

Johnsson, R., Axelsson, P., and Stromqvist, B. (1997). Posterolateral lumbar fusion using facet joint fixation with biodegradable rods: a pilot study. *Eur Spine J*, 6, 144-148.

Kandziora, F., Pflugmacher, R., Kleemann, R., Duda, G., Wise, D. L., Trantolo, D. J., and Lewandrowski, K. U. (2002). Biomechanical analysis of biodegradable interbody fusion cages augmented With poly(propylene glycol-co-fumaric acid). *Spine*, 27, 1644-1651.

Kandziora, F., Pflugmacher, R., Scholz, M., Eindorf, T., Schnake, K. J., and Haas, N. P. (2004). Bioabsorbable interbody cages in a sheep cervical spine fusion model. *Spine*, 29, 1845-1855.

Khodadadyan-Klostermann, C., Kandziora, F., Schnake, K. J., Lewandrowski, K. U., Wise, D., Weiler, A., and Haas, N. P. (2001). [Mechanical comparison of biodegradable intervertebral lumbar cages]. *Chirurg*, 72, 1431-1438.

Klopp, L. S., Welch, W. C., Tai, J. W., Toth, J. M., Cornwall, G. B., and Turner, A. S. (2004). Use of polylactide resorbable film as a barrier to postoperative peridural adhesion in an ovine dorsal laminectomy model. *Neurosurg Focus*, 16, E2.

Krijnen, M. R., Smit, T. H., Strijkers, G. J., Nicolay, K., Pouwels, P. J., and Wuisman, P. I. (2004). The use of high-resolution magnetic resonance imaging for monitoring interbody fusion and bioabsorbable cages: an ex vivo pilot study. *Neurosurg Focus*, 16, E3.

Krijnen, M. R., Mullender, M. M., Smit, T. H., Everts, V., and Wuisman, P.I.J.M. Radiographical, histological and chemical evaluation of bioresorbable 70/30 poly (L-lactide-CO-D,L-lactide) interbody fusion cages in a goat model. *Spine*, (2006) in press.

Kuklo, T. R., Rosner, M. K., and Polly, D. W., Jr. (2004). Computerized tomography evaluation of a resorbable implant after transforaminal lumbar interbody fusion. *Neurosurg Focus*, 16, E10.

Kulkarni, R. K., Pani, K. C., Neuman, C., and Leonard, F. (1966). Polylactic acid for surgical implants. *Arch Surg*, 93, 839-843.

Lanman, T. H. and Hopkins, T. J. (2004a). Early findings in a pilot study of anterior cervical interbody fusion in which recombinant human bone morphogenetic protein-2 was used with poly(L-lactide-co-D,L-lactide) bioabsorbable implants. *Neurosurg Focus*, 16, E6.

Lanman, T. H. and Hopkins, T. J. (2004b). Lumbar interbody fusion after treatment with recombinant human bone morphogenetic protein-2 added to poly(L-lactide-co-D,L-lactide) bioresorbable implants. *Neurosurg Focus*, 16, E9.

Leclercq, T. A. (2001). Complication of cages. *J Neurosurg*, 95, 283.

Leenslag, J. W., Pennings, A. J., Bos, R. R., Rozema, F. R., and Boering, G. (1987). Resorbable materials of poly(L-lactide). VII. In vivo and in vitro degradation. *Biomaterials*, 8, 311-314.

Li, S. (1999). Hydrolytic degradation characteristics of aliphatic polyesters derived from lactic and glycolic acids. *J Biomed Mater Res*, 48, 342-353.

Li, S. and McCarthy, S. (1999). Further investigations on the hydrolytic degradation of poly (DL-lactide). *Biomaterials*, 20, 35-44.

Lippman, C. R., Hajjar, M., Abshire, B., Martin, G., Engelman, R. W., and Cahill, D. W. (2004). Cervical spine fusion with bioabsorbable cages. *Neurosurg Focus*, 16, E4.

Loo, J. S., Ooi, C. P., and Boey, F. Y. (2005a). Degradation of poly(lactide-co-glycolide) (PLGA) and poly(L-lactide) (PLLA) by electron beam radiation. *Biomaterial*s, 26, 1359-1367.

Loo, S. C., Ooi, C. P., and Boey, Y. C. (2005b). Influence of electron-beam radiation on the hydrolytic degradation behaviour of poly(lactide-co-glycolide) (PLGA). *Biomaterials*, 26, 3809-3817.

Loo, S. C., Tan, H. T., Ooi, C. P., and Boey, Y. C. (2006). Hydrolytic degradation of electron beam irradiated high molecular weight and non-irradiated moderate molecular weight PLLA. *Acta Biomater*, 2, 287-296.

Lowe, T. G. and Coe, J. D. (2002a). Bioresorbable polymer implants in the unilateral transforaminal lumbar interbody fusion procedure. *Orthopedics*, 25, s1179-s1183.

Lowe, T. G. and Coe, J. D. (2002b). Resorbable polymer implants in unilateral transforaminal lumbar interbody fusion. *J Neurosurg*, 97, 464-467.

Mainil-Varlet, P., Curtis, R., and Gogolewski, S. (1997a). Effect of in vivo and in vitro degradation on molecular and mechanical properties of various low-molecular-weight polylactides. *J Biomed Mater Res*, 36, 360-380.

Mainil-Varlet, P., Rahn, B., and Gogolewski, S. (1997b). Long-term in vivo degradation and bone reaction to various polylactides. 1. One-year results. *Biomaterials*, 18, 257-266.

Matsumoto, M., Chosa, E., Nabeshima, K., Shikinami, Y., and Tajima, N. (2002). Influence of bioresorbable, unsintered hydroxyapatite/poly-L-lactide composite films on spinal cord, nerve roots, and epidural space. *J Biomed Mater Res*, 60, 101-109.

Matsusue, Y., Hanafusa, S., Yamamuro, T., Shikinami, Y., and Ikada, Y. (1995). Tissue reaction of bioabsorbable ultra high strength poly (L-lactide) rod. A long-term study in rabbits. *Clin Orthop Relat Res*, 246-253.

Matthews, I. P., Gibson, C., and Samuel, A. H. (1989). Enhancement of the kinetics of the aeration of ethylene oxide sterilized polymers using microwave radiation. J *Biomed.Mater Res*. 23, 143-156.

McAfee, P. C., Cunningham, B. W., Lee, G. A., Orbegoso, C. M., Haggerty, C. J., Fedder, I. L., and Griffith, S. L. (1999). Revision strategies for salvaging or improving failed cylindrical cages. *Spine,* 24, 2147-2153.

Middleton, J. C. and Tipton, A. J. (2000). Synthetic biodegradable polymers as orthopedic devices. *Biomaterials*, 21, 2335-2346.

Naderi, S. and Arda, N. (2001). Complication of cages. *J Neurosurg*, 95, 282-283.

Nuutinen, J. P., Clerc, C., Virta, T., and Tormala, P. (2002). Effect of gamma, ethylene oxide, electron beam, and plasma sterilization on the behaviour of SR-PLLA fibres in vitro. *J Biomater.Sci Polym Ed*. 13, 1325-1336.

Oudega, M., Gautier, S. E., Chapon, P., Fragoso, M., Bates, M. L., Parel, J. M., and Bunge, M. B. (2001). Axonal regeneration into Schwann cell grafts within resorbable poly(alpha-hydroxyacid) guidance channels in the adult rat spinal cord. *Biomaterials*, 22, 1125-1136.

Patist, C. M., Mulder, M. B., Gautier, S. E., Maquet, V., Jerome, R., and Oudega, M. (2004). Freeze-dried poly(D,L-lactic acid) macroporous guidance scaffolds impregnated with brain-derived neurotrophic factor in the transected adult rat thoracic spinal cord. *Biomaterials*, 25, 1569-1582.

Pflugmacher, R., Schleicher, P., Gumnior, S., Turan, O., Scholz, M., Eindorf, T., Haas, N. P., and Kandziora, F. (2004). Biomechanical comparison of bioabsorbable cervical spine interbody fusion cages. *Spine*, 29, 1717-1722.

Pietrzak, W. S., Sarver, D., and Verstynen, M. (1996). Bioresorbable implants--practical considerations. *Bone*, 19, 109S-119S.

Pietrzak, W. S., Sarver, D. R., and Verstynen, M. L. (1997). Bioabsorbable polymer science for the practicing surgeon. *J Craniofac Surg*, 8, 87-91.

Poynton, A. R., Tomin, E., Cornwall, G. B., Zheng, F., and Lane, J. M. (2002a). Resorbable posterolateral graft containment in a rabbit spinal fusion model. *Orthopedic*s, 25, s1173-s1177.

Poynton, A. R., Zheng, F., Tomin, E., Lane, J. M., and Cornwall, G. B. (2002b). Resorbable posterolateral graft containment in a rabbit spinal fusion model. *J Neurosurg*, 97, 460-463.

Santos, E. R., Goss, D. G., Morcom, R. K., and Fraser, R. D. (2003). Radiologic assessment of interbody fusion using carbon fiber cages. *Spine*, 28, 997-1001.

Shikinami, Y. and Okuno, M. (2003). Mechanical evaluation of novel spinal interbody fusion cages made of bioactive, resorbable composites. *Biomaterials*, 24, 3161-3170.

Smit, T. H., Krijnen, M. R., van Dijk, M., and Wuisman, P. Application of polylactides in spinal cages: Studies in goat model. *J Mater Sci Mater Med*, in press. 2006a.

Smit, T. H., Thomas, K. A., Hoogendoorn, R. J. W., Strijkers, G. J., Helder, M. N., and Wuisman, P. I. Sterilization and strength of 70/30 polylactide cages: e-beam vs Ethylene oxide. unpublished data . 2006b.

Smit, T. H. (2002). The use of a quadruped as an in vivo model for the study of the spine - biomechanical considerations. *Eur Spine J*, 11, 137-144.

Smit, T. H., Muller, R., van Dijk, M., and Wuisman, P. I. (2003). Changes in bone architecture during spinal fusion: three years follow-up and the role of cage stiffness. *Spine*, 28, 1802-1808.

Soderlund, C. H., Pointillart, V., Pedram, M., Andrault, G., and Vital, J. M. (2004). Radiolucent cage for cervical vertebral reconstruction: a prospective study of 17 cases with 2-year minimum follow-up. *Eur.Spine J.* 13, 685-690.

Sturm, S., Engels, T., Smit, T. H., Govaert, L. E. (2006). Premature failure of poly-L/D-lactide bioresorbable cages: pitfalls in designing time dependent materials. 13th international conference deformation, yield and fracture of polymers, 315-318.

Togawa, D., Bauer, T. W., Lieberman, I. H., Lowery, G. L., and Takikawa, S. (2003). Histology of tissues within retrieved human titanium mesh cages. *Spine,* 28, 246-253.

Togawa, D., Bauer, T. W., Lieberman, I. H., and Sakai, H. (2004). Lumbar intervertebral body fusion cages: histological evaluation of clinically failed cages retrieved from humans. *J Bone Joint Surg Am*, 86-A, 70-79.

Toth, J. M., Estes, B. T., Wang, M., Seim, H. B., III, Scifert, J. L., Turner, A. S., and Cornwall, G. B. (2002). Evaluation of 70/30 poly (L-lactide-co-D,L-lactide) for use as a resorbable interbody fusion cage. *J Neurosurg Spine*, 97, 423-432.

Totoribe, K., Matsumoto, M., Goel, V. K., Yang, S. J., Tajima, N., and Shikinami, Y. (2003). Comparative biomechanical analysis of a cervical cage made of an unsintered hydroxyapatite particle and poly-L-lactide composite in a cadaver model. *Spine*, 28, 1010-1015.

Tullberg, T. (1998). Failure of a carbon fiber implant. A case report. *Spine*, 23, 1804-1806.

Tunc, D. C., van Dijk, M., Smit, T., Higham, P., Burger, E., and Wuisman, P. (2004). Three-year follow-up of bioabsorbable PLLA cages for lumbar interbody fusion: in vitro and in vivo degradation. *Adv Exp Med Biol*, 553, 243-255.

van der, Elst. M., Klein, C. P., Blieck-Hogervorst, J. M., Patka, P., and Haarman, H. J. (1999). Bone tissue response to biodegradable polymers used for intra medullary fracture fixation: a long-term in vivo study in sheep femora. *Biomaterials*. 20, 121-128.

Vaccaro, A. R., Robbins, M. M., Madigan, L., Albert, T. J., Smith, W., and Hilibrand, A. S. (2004). Early findings in a pilot study of anterior cervical fusion in which bioabsorbable interbody spacers were used in the treatment of cervical degenerative disease. *Neurosurg.Focus*. 16, E7.

van Dijk, M., Smit, T. H., Arnoe, M. F., Burger, E. H., and Wuisman, P. I. (2003). The use of poly-L-lactic acid in lumbar interbody cages: design and biomechanical evaluation in vitro. *Eur Spine J*, 12, 34-40.

van Dijk, M., Smit, T. H., Burger, E. H., and Wuisman, P. I. (2002a). Bioabsorbable poly-L-lactic acid cages for lumbar interbody fusion: three-year follow-up radiographic, histologic, and histomorphometric analysis in goats. *Spine*, 27, 2706-2714.

van Dijk, M., Smit, T. H., Burger, E. H., and Wuisman, P. I. (2002b). Bone histomorphometric evaluation of a clinically fused titanium tumour cage in a child. Eur *Spine J*, 11, 507-511.

van Dijk, M., Smit, T. H., Sugihara, S., Burger, E. H., and Wuisman, P. I. (2002c). The effect of cage stiffness on the rate of lumbar interbody fusion: an in vivo model using poly(l-lactic Acid) and titanium cages. *Spine*, 27, 682-688.

van Dijk, M., Tunc, D. C., Smit, T. H., Higham, P., Burger, E. H., and Wuisman, P. I. (2002d). In vitro and in vivo degradation of bioabsorbable PLLA spinal fusion cages. J *Biomed Mater Res*, 63, 752-759.

van Dijk, M., van Diest, P. J., Smit, T. H., Berkhof, H., Burger, E. H., and Wuisman, P. I. (2005). Four-year follow-up of poly-L-lactic Acid cages for lumbar interbody fusion in goats. *J Long Term Eff Med Implants*, 15, 125-138.

Weir, N. A., Buchanan, F. J., Orr, J. F., and Dickson, G. R. (2004a). Degradation of poly-L-lactide. Part 1: in vitro and in vivo physiological temperature degradation. *Proc Inst Mech Eng* [H], 218, 307-319.

Weir, N. A., Buchanan, F. J., Orr, J. F., Farrar, D. F., and Boyd, A. (2004b). Processing, annealing and sterilisation of poly-L-lactide. *Biomaterials*, 25, 3939-3949.

Weir, N. A., Buchanan, F. J., Orr, J. F., Farrar, D. F., and Dickson, G. R. (2004c). Degradation of poly-L-lactide. Part 2: increased temperature accelerated degradation. *Proc Inst Mech Eng* [H], 218, 321-330.

Welch, W. C., Thomas, K. A., Cornwall, G. B., Gerszten, P. C., Toth, J. M., Nemoto, E. M., and Turner, A. S. (2002). Use of polylactide resorbable film as an adhesion barrier. *J Neurosurg*, 97, 413-422.

Williams, D. F. (1992). Mechanisms of biodegradation of implantable polymers. *Clin Mater*, 10, 9-12.

Wuisman, P. I. and Smit, T. H. (2006). Bioresorbable polymers: heading for a new generation of spinal cages. *Eur Spine J*, 15, 133-148.

Wuisman, P. I., van Dijk, M., and Smit, T. H. (2002). Resorbable cages for spinal fusion: an experimental goat model. *Orthopedics*, 25, s1141-s1148.

Zhou, S., Causey, T. B., Hasona, A., Shanmugam, K. T., and Ingram, L. O. (2003). Production of optically pure D-lactic acid in mineral salts medium by metabolically engineered Escherichia coli W3110. *Appl Environ Microbiol*, 69, 399-407.

In: Biomaterials Research Advances
Editor: J. B. Kendall, pp. 93-143
ISBN: 978-1-60021-892-7

Chapter 5

Nanocrystalline Apatite-Based Biomaterials: Synthesis, Processing and Characterization

D. Eichert, C. Drouet, H. Sfihi[a, b], C. Rey and C. Combes
CIRIMAT UMR UPS-INPT-CNRS 5085, Equipe Physico-Chimie des Phosphates, ENSIACET, 118 route de Narbonne, 31077 Toulouse cedex 4, France
[a]Laboratoire de Physique Quantique, UMR CNRS 7142, ESPCI, 10 rue Vauquelin, 75231 Paris, France
[b]Département de Physique, UFR SMBH, Université Paris 13, 74 rue Marcel Cachin, 93102 Bobigny Cedex, France

Abstract

The improvement of the biological activity and performance of bone substitute materials is one of the main concerns of orthopaedic and dental surgery specialists. Biomimetic nanocrystalline apatites exhibit enhanced and tunable reactivity as well as original surface properties related to their composition and mode of formation. Synthetic nanocrystalline apatites analogous to bone mineral can be easily prepared in aqueous media and one of their most interesting characteristics is the existence of a hydrated surface layer containing labile ionic species. Ion exchange and macromolecule adsorption processes can easily and rapidly take place due to strong interactions with the surrounding fluids. The ion mobility in the hydrated layer allows direct crystal-crystal or crystal-substrate bonding. The fine characterization of these very reactive nanocrystals is essential and can be accomplished with different tools including chemical analysis and spectroscopic techniques such as FTIR, Raman and solid state NMR. The reactivity of the hydrated layer of apatite nanocrystals offers material scientists and medical engineers extensive possibilities for the design of biomaterials with improved bioactivity using unconventional processing. Indeed apatitic biomaterials can be processed at low temperature which preserves their surface reactivity and biological properties. They can also be associated in various ways with active molecules and/or ions. Several examples of use and processing of nanocrystalline apatites involved in the preparation of tissue-engineered biomaterials, cements, ceramics, composites and coatings on metal prostheses are presented.

INTRODUCTION

Osteoarticular pathologies are, at all ages, the first cause of handicap and raise concern in public healthcare. Articular aging, traumatology, child growth defects, bone tumor treatment, osteoporosis and related bone failures, constitute points for which research efforts are necessary. Bone diseases and induced defects are also of prime importance in maxillo-facial surgery and odontology especially for aging populations. The shortcomings of bone auto- or allografts which in addition involve secondary operations, risks of disease transmission as well as immunological rejection and morbidity justified the development of synthetic bone graft materials.

In the past decades, the effort to adapt the first biomaterials taken from other technical domains (e.g. alumina, carbon, titanium carbide and nitride, plaster of Paris), or to design new materials better suited to biological applications, led to significant advances. However in the opinion of many researchers the ultimate achievement would be the perfect imitation of biological tissues and more importantly the improvement of biological repair and maintenance processes. Ideally, a substitute material should mimic the living tissue's mechanical, chemical, biological and functional properties; however the design of a complex structure such as bone is still impossible to achieve without the aid of Nature which masters, using sophisticated chemical properties and processes, this high performance mineral-protein composite.

Poorly crystalline apatites (PCA) are the major inorganic constituent of mineralized tissues in vertebrates. The imitation of bone mineral has inspired the research and development of calcium phosphate (CaP) based biomaterials. The most famous and most widely used CaP compound is stoichiometric hydroxyapatite processed as dense or porous ceramics, coatings and composites. However, the design of biomaterials has evolved and today the function of a biomaterial is not restricted to physical substitution but the novel biomaterial should actively participate in the process of bone regeneration implying reactivity of the component(s). In this view we will show in this chapter how non-stoichiometric nanocrystalline apatite-based biomaterials can fulfil two of the main challenges: mimicking bone mineral crystal structure and composition and exhibiting a controlled reactivity regarding interactions with components of biological fluids (ions, proteins).

This chapter reports part of our "bioinspired" research based on the idea that the development and the processing of bioactive biomaterials for bone substitution or regeneration applications can take advantage of a thorough knowledge and understanding of the structure and properties of the tissue to be substituted. Due to the complex structure and heterogeneity of biological systems such as bone, synthetic apatites are generally used to throw light on the surface reactivity of bone mineral. However the characterization of nanocrystalline apatite analogous to bone mineral is difficult due to its relative instability and poor crystallinity. Even though the results are complex, recent fine investigations on synthetic PCA revealing their original surface reactivity are useful both in biomaterials or biomineralization.

The present chapter gathers the main results of our most recent investigations on nanocrystalline apatite composition, structure and properties with the aim of better understanding bone mineral properties and achieving better design of new bioactive nanocrystalline apatite based biomaterials or improving the biological performance of existing bone substitutes. This chapter which alternates between results on synthetic apatites and their significance for biological apatites is organized in seven sub-sections: 1) a description of bone composition and structure and the way it was perceived through the 20th century are presented first, 2) synthetic routes of nanocrystalline apatites are discussed with an emphasis on biomimetic nanocrystalline apatite preparation and maturation at room temperature and at physiological pH as developed in our research group, 3) the complementary characterization techniques that we use to investigate the global and local fine structure and composition of nanocrystalline apatites are reported 4) a model for apatite nanocrystals is put forward which can explain the properties of synthetic and biological apatite nanocrystals, 5) the physico-chemical properties of biomimetic apatites and their involvement in the biological behavior of PCA based materials are presented and discussed, 6) examples of nanocrystalline apatite based biomaterial showing some potential with regards to nanocrystalline apatite surface reactivity are described, 7) finally, the biological properties of apatites are presented.

1. Early Works and the Way Bone Mineral was Conceived

The studies on bone mineral composition and structure were rather puzzling for the first investigators. Before the development of structural analysis by X-ray diffraction (XRD), until the beginning of the 20th century, chemical composition was one of the major characterization tools used to identify biominerals. Other identification methods, for example based on the use of the optical properties of crystals could not be applied to biological apatites due to the small size of their crystals. Chemical analyses revealed the diversity of phosphate-containing biominerals. Three major components are always present: calcium, phosphate and carbonate, and were readily identified but they showed variable contents depending on the species, the individuals, their location in the body or the age and the type of mineralized tissue considered, in contrast with the other major biomineral, calcium carbonate, showing a rather constant composition. It was then accepted, in accordance with the hypothesis of Haüy concerning the composition of calcium phosphate-carbonate minerals, that several phases co-existed in hard tissues of vertebrates: essentially tricalcium phosphate and calcium carbonate [McConnel 1973].

The first structural identifications of biominerals using X-ray diffraction were obtained by de Jong in 1926 [de Jong 1926]. He established that calcium phosphate biominerals of vertebrates corresponded to an apatite structure and since that time bone mineral has been frequently identified as hydroxyapatite (HA):

$$Ca_{10}(PO_4)_6(OH)_2$$

which was later considered to crystallize in the hexagonal system (space group $P6_3/m$), see on figure 1, and then shown to be monoclinic, at room temperature, when stoichiometric [Posner

1958, Elliott 1973]. The XRD data revealed very broad bands indicating low cristallinity. However, the identification of a poorly crystalline apatite structure did not explain the existence of significant amounts of carbonate in all mineralized biological tissues. Thus it was considered that bone mineral was a mixture of poorly crystalline apatite and amorphous calcium carbonate or that carbonate ions were at least in part adsorbed on the apatite crystal surface [Elliott 1994].

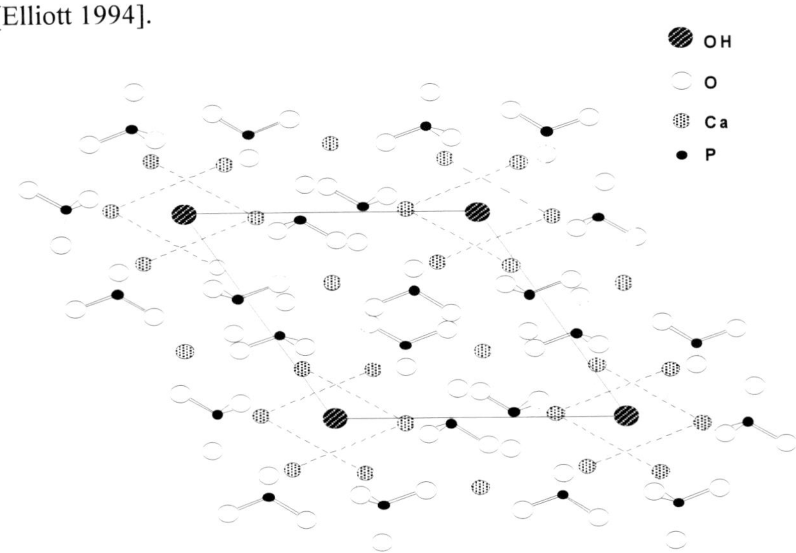

Figure 1. Projection on the (001) plane of hydroxyapatite structure. The two Ca^{2+} triangles lining the "tunnels" of the structure are located at z ¼ and ¾. OH^- ions are slightly under or above the triangles [Cazalbou 2004] - Reproduced by permission of The Royal Society of Chemistry.

It was only in the sixties that studies on synthetic carbonated apatites established that all carbonate ions could in fact be located within the apatite structure [Legeros 1968, Labarthe 1973]. But this has been the object of much controversy. Detailed studies indicated that carbonate ions could be located in the two anionic sites of the apatite structure: in PO_4^{3-} sites (type B carbonated apatite) and OH^- sites (type A carbonated apatite). Bone apatites were believed to correspond essentially to type B carbonated apatite whereas enamel contained both type A and B carbonate. The fraction of type A carbonate in dental enamel was evaluated at about 10% of the total carbonate content using infrared spectroscopy [Elliott 1985]. The use of carbonated apatite as a model for biological calcifications of hard tissues of vertebrates is nowadays accepted. The variability of the composition of apatite minerals and their mode of formation however needed further investigation.

From a kinetic point of view the direct formation of apatite crystals in calcified tissues was considered as unlikely, regarding the slow growth rate of apatite crystals and the existence in body fluids of crystal growth inhibitors such as magnesium and carbonate ions [Campbell 1991]. The improvement of crystallinity of bone apatites upon aging, revived for a while the theory of an amorphous phase considered as a necessary precursor in the formation of biological apatites [Termine 1966]. This hypothesis based on studies dealing with the formation and stability of amorphous calcium phosphates and its progressive conversion into apatite, compared to bone mineral evolution upon aging, was thus considered as quite consistent for a decade. In the eighties however it was demonstrated that bone mineral was composed of apatite nanocrystals with no or a non-detectable amorphous phase [Grynpas 1984]. The peculiar shape of bone crystals (platelets) showing non-equivalent a and b directions perpendicular to the c-axis of the hexagonal structure (figure 1), however led to the hypothesis that apatite formation involved another precursor phase, very close to apatite and indiscernible considering the very poor crystallization state of the mineral: triclinic OctaCalcium Phosphate (OCP) [Brown 1987]:

$$Ca_8\,(PO_4)_4\,(HPO_4)_2,\ 5H_2O$$

The OCP structure has been shown to consist in the association of an apatite-like layer and a hydrated layer [Mathew 1988]. Precipitating as platelet-shaped crystals, this phase exhibits a high crystal growth rate and hydrolyzes readily in aqueous media into apatite, forming interlayered compounds with hydroxyapatite and preserving the original platelet shape of the crystals. This model does not however explain all the variability of apatite compositions and particularly the very similar role played by carbonate and HPO_4^{2-} ions in bone mineral and synthetic analogues [Neuman 1956].

The chemical composition of biological apatite has been the object of several approximations frequently based on the composition of model minerals or synthetic analogues. A general chemical formula proposed by Winand for HPO_4^{2-}-containing apatite was [Winand 1961]:

$$Ca_{10-x}\,(PO_4)_{6-x}\,(HPO_4)_x\,(OH)_{2-x} \text{ with } 0 \leq x \leq 2$$

and by Labarthe et al. for carbonate-containing apatites [Labarthe 1973]:

$$Ca_{10-x}\,(PO_4)_{6-x}\,(CO_3)_x\,(OH)_{2-x} \text{ with } 0 \leq x \leq 2.$$

These formula establish a similar behavior for bivalent ion substitution of trivalent phosphates: the creation of a cationic vacancy and an anionic vacancy in monovalent sites. These chemical formulas are consistent with the limit composition observed (x=2) and the decrease of the OH^- content when the amount of carbonate and/or HPO_4^{2-} in the apatite increases. Other chemical formulas have been proposed. The most general one [Rey 2006]:

$$Ca_{10-x+u}\,(PO_4)_{6-x-y}\,(HPO_4^{2-} \text{ or } CO_3^{2-})_{x+y}\,(OH)_{2-x+2u-y} \text{ with } 0 \leq x \leq 2 \text{ and } 0 \leq 2u + y \leq x$$

is however of little relevance for biological apatites which are best approximated by the simple combination of the two previous formulas taking into account the possible existence of type A carbonates:

$$Ca_{10-x}\,(PO_4)_{6-x}\,(HPO_4 \text{ or } CO_3)_x\,(OH \text{ or } \tfrac{1}{2}\,CO_3)_{2-x} \text{ with } 0 \leq x \leq 2$$

The compilation of different cortical bone analyses suggests a relatively homogeneous composition [Legros 1987]:

$$Ca_{8.3}\,(PO_4)_{4.3}\,(HPO_4 \text{ or } CO_3)_{1.7}\,(OH \text{ or } \tfrac{1}{2}\,CO_3)_{0.3}$$

characterized by a very high vacancy content close to the maximum (x=1.7). The carbonate content varies with age: it is very low in embryonic bone, and can represent up to 80% of the bivalent ions in the bone mineral of old vertebrate animals. The OH^- content of bone is very low at any age and OH^- ions can barely be detected [Rey 1995, Pasteris 2004]. The composition of tooth enamel crystals reveals a radically different chemical composition:

$$Ca_{9.4}\ (PO_4)_{5.4}\ (HPO_4 \text{ or } CO_3)_{0.6}\ (OH \text{ or } \tfrac{1}{2}\ CO_3)_{1.4}$$

showing a much lower vacancy content, unveiling the unique adaptability of apatites to their biological functions [Cazalbou 2004].

Other minor substitutions are found in biological apatites involving for example trivalent cations (e.g. rare earth elements, actinides) or monovalent cations (especially Na^+) for Ca^{2+}, tetravalent ions replacing PO_4^{3-}, and bivalent ions replacing OH^-. Several charge compensation mechanisms have been proposed. Although the ability of the apatite structure to fix many elements has several consequences regarding intoxications with mineral ions and diseases, such possibilities seem to have a minor influence on the chemical formula of apatites in calcified tissues due to the low amounts of these foreign ions [Iyengar 1999].

The present data underline the strong heterogeneity of bone mineral and apatitic biomineralizations. The global compositions do not reflect strong local variations between osteons and within osteons, and probably between the crystals themselves [Paschalis 1996]. In addition several properties of bone mineral such as ion exchange suggest the existence of surface modifications and possibly surface compositions different from the bulk at the level of a nanocrystal. The heterogeneities of the mineral are among its chief characteristics, mainly related to bone remodeling processes and formation conditions. Parameters susceptible to evaluate these characteristics would be of great utility.

The replacement and healing of damaged hard tissues have always been a concern for human beings as shown by the examination of mummies. It is however only very recently that calcium phosphates have been used for bone substitution and repair [Jarcho 1979]. The first to be used were stoichiometric hydroxyapatite (HA) and β-tricalcium phosphate (β-TCP) which are stable CaP at high temperature and can be easily sintered into ceramics. They are still the major industrial CaP biomaterials. β-TCP was shown to be bioabsorbable and replaced by bone whereas HA constituted non-degradable materials. β-TCP is mainly used as a bioceramic whereas HA is also being processed for other biomaterials uses such as the coating of metallic prostheses where it was found to considerably improve bone repair as an "osteoconductive" material or composite ceramic-polymer materials showing strong mechanical analogies with bone tissues and excellent bone bonding abilities [de Groot 1987, Bonfield 1988]. Biphasic Calcium Phosphates (BCP), associating these two high-temperature CaP allow a controlled resorption rate and have been reported to offer superior biological properties [Daculsi 2003, Legeros 2002]. They are progressively replacing β-TCP ceramics in Europe. A new technological step was made with the development of CaP cements [Brown 1986]. These materials are able to set and harden in a living body and most can be injected. Despite their poor mechanical properties they offer a number of advantages and are increasingly used for several applications. More recently biomimetic coatings involving low temperature nanocrystalline CaP have been proposed - some have been claimed to exhibit osteoinductive properties [Habibovic 2006].

2. Synthesis of Nanocrystalline Apatites

As mentioned above, synthetic nanocrystalline apatites are of undeniable interest in the preparation of apatite-based bioceramics for bone substitution, repair or augmentation applications. However, synthetic apatites are also prepared and studied to better understand the formation of biological apatites and some of their properties [Legeros 1994]. Since the eighties, several synthetic routes have emerged and in the near future they could challenge the high-energy conventional processes involving high temperatures. Among the major advantages of these emerging unconventional processes are the use of low temperature (from room temperature (RT) to about 400°C), the flexibility and range of chemical compositions, and the physical, chemical and biological properties of these CaP.

Synthetic apatites can be prepared by several methods (precipitation under conditions of constant or changing composition, hydrolysis, solid/solid reaction at high temperature, hydrothermal methods) the type of which determines the amount and kind of substitution in the apatite. In this section we will focus on the synthesis of calcium-deficient apatites in solution systems.

Several processes (precipitation by double decomposition, sol-gel method, hydrolysis) involving various media (aqueous, hydro-alcoholic or organic solutions) leading to calcium phosphate apatites have been reported. Sol-gel processes still raise some problems: the long time needed for the preparation of the sol, and the presence of other calcium phosphate phases depending on aging time and temperature [Liu 2002]. Calcium-deficient or substituted apatites can also be prepared by hydrolysis of amorphous calcium phosphate, dicalcium phosphate dihydrate, octacalcium phosphate, or α and β tricalcium phosphate for example. Hydrolysis of these calcium phosphate phases to yield apatite can proceed through a dissolution-reprecipitation mechanism depending on the pH, the temperature and the presence of other ions. The latter can act as inhibitors (magnesium, pyrophosphate ions) or promotors (fluoride ions) of hydrolysis of dicalcium phosphate dihydrate (DCPD: $CaHPO_4\ 2H_2O$) and OCP for example [Legeros 1994].

Two major parameters determine the crystallinity and the calcium deficiency of the apatite obtained by precipitation methods: temperature (ambient temperature to 100°C) and pH (basic). When precipitated from solutions at temperatures between 80°C and 100°C, the higher the initial pH, the lower the calcium deficiency. Precipitation at temperatures under 80°C leads to less and less crystallized apatites, and the synthesis of poorly crystalline apatites analogous to bone mineral can be easily achieved at ambient temperature and physiological pH according to the method reported in the next sub-section.

Interestingly, precipitation using a hydro-alcoholic medium with a dielectric constant lower than that of water, provides control of hydrogenphosphate and carbonate ion content in apatite analogous to bone mineral [Zahidi 1985, Rodrigues 1998, Dabbarh 2000]. In addition, several studies reported the influence of the drying process and temperature on the composition, structure and degree of crystallinity of apatitic calcium phosphates [Dabbarh 2000, Lebugle 1986].

2.A. Poorly Crystalline Apatite (PCA) Synthesis

The results presented in this chapter are related to poorly crystalline apatites synthesized at ambient temperature and physiological pH by double decomposition between a phosphate and carbonate solution (for example, 40g of $(NH_4)_2HPO_4$, 20g of $NaHCO_3$ and concentrated ammonia solution 1 ml in 500 ml of deionized water) and a calcium solution (fr example, 17.7g of $Ca(NO_3)_2$ $4H_2O$ in 250 ml of deionized water) as previously published [Rey 1989]. The calcium solution is rapidly poured into the phosphate and carbonate solution at room temperature (20°C) and stirred only for a few minutes. For investigations on freshly-precipitated nanocrystalline apatites, the precipitate is then very quickly filtered under vacuum and washed with deionized water (2 liters). Then the gel is freeze-dried and finally stored in a freezer to prevent further maturation of the PCA nanocrystals. This method leads to a carbonated poorly crystalline apatite analogous to bone mineral [Rey 1995]. Non-carbonated poorly crystalline apatite can be prepared by this method with a carbonate-free phosphate solution. Other recipes involving cationic and anionic solutions with slightly different Ca/P ratio and no concentrated ammonia solution also leads to poorly crystalline apatite. In all cases, the large excess of phosphate (and bicarbonate) ions in the solution provides pH buffering at pH = 7.4.

2.B. Poorly Crystalline Apatite (PCA) Maturation

To study the physical-chemical properties of nanocrystalline apatites, the apatite can be left to mature after precipitation at room temperature in the mother solution without stirring and in a stoppered vial to minimize the release and uptake of CO_2 at physiological pH. This evolution in solution (maturation) is an important process that can help us understand the evolution of the composition, structure and properties of biological and synthetic biomimetic apatites after different aging times (corresponding to young and old bones for example).

After maturation during variable periods of time (from an hour to several months), the precipitates were filtered under vacuum and washed with deionized water. In the case of subsequent ion exchange experiments, part of the gel is then freeze-dried (reference sample) and the other part is used for ion exchange treatments.

The study of maturation properties of PCA is presented in section 5.A.

2.C. Ionic Exchange (Direct and Inverse) on Poorly Crystalline Apatite (PCA)

Two kinds of ion exchange experiments can be performed on PCA: anionic exchange (and the study of the reversibility of exchange between carbonate and hydrogenophosphate) ions, and cationic exchange (and the study of the reversibility of exchange between calcium and other cations such as strontium or magnesium ions as reported in this chapter). Also, such ion exchange experiments can be carried out on immature gels or on apatite samples matured for various durations.

The ion exchange is performed by exposing either the gel or the mature sample of nanocrystalline carbonated apatite (CA) to an "exchange" solution containing the target ion (HCO_3^- or HPO_4^{2-} for anionic hydrogenphosphate ⇔ carbonate exchange, Mg^{2+} or Sr^{2+} for cationic exchange with Ca^{2+}) at varying concentrations (for example 1 M for 10 minutes). The starting salts used for the preparation of the "exchange" solution can be $NaHCO_3$, $(NH_4)_2HPO_4$, $Mg(NO_3)_2$ and $Sr(NO_3)_2$. Part of the "exchanged" samples are then filtered, washed with deionized water and freeze-dried. The other part is re-suspended for 10 minutes in the "inverse exchange" solution, containing the initial ion at a concentration of 1 M. In all cases, the samples are finally filtered, washed with deionized water and freeze-dried. In the text the notation "CA/X" represents a carbonated apatite exchanged with the ion X, and "CA/X/Y" represents the same sample after inverse exchange with the ion Y. For example, "CA/Sr" refers to a carbonated apatite for which part of the calcium ions has been exchanged with strontium ions, and "CA/Sr/Ca" refers to the same sample after inverse exchange of Sr^{2+} by Ca^{2+}.

The studies of ionic exchange properties of PCA are presented in section 5.B.

3. Characterization of Apatites

3.A. Chemical Analysis

Depending on the precipitation conditions, on the maturation time and/or on ion exchange treatments, the composition of calcium-deficient apatites can vary significantly. The determination of calcium, total phosphate and carbonate ions can be easily performed in different ways whereas the direct evaluation of hydrogenphosphate ions which are one of the most important markers of PCA and bone mineral crystals has not yet been possible (only indirect measurements are available). All the chemical analysis methods used for apatite characterization are based on the dissolution of apatite in acidic solution before the analysis (calcium and orthophosphate ions determination) or during the analysis (carbonate ions).

Calcium concentration can be determined by complexometry with EDTA and the phosphorus concentration by UV-visible spectrophotometry of the phospho-vanado-molybdenum complex. The Ca/P atomic ratio of apatites can be calculated from the result of these two analyses. The relative uncertainty on calcium and phosphorus concentrations has been evaluated at 0.5 %.

The titration of HPO_4^{2-} ions is more complex because the two inorganic orthophosphate ions, PO_4^{3-} and HPO_4^{2-}, encountered in apatites are in rapid equilibrium in solution. So the dissolution of apatite in acidic solution and the determination of total phosphorus content by colorimetry cannot distinguish HPO_4^{2-} and PO_4^{3-}. Therefore, the only way to determine the HPO_4^{2-} content is to condense the ions into pyrophosphates according to equation 1. Then the HPO_4^{2-} level is determined by chemical analysis using the Gee and Dietz method which is based on the formation of pyrophosphate in apatite-containing HPO_4^{2-} ions upon heating [Gee 1953]. During treatment of the apatite at 500°C for 3 hours (or 600°C for 20 min) the following reaction occurs leading to the formation of pyrophosphate ions [Gee 1955]:

$$2\ HPO_4^{2-} \rightarrow P_2O_7^{4-} + H_2O \qquad \text{(eq. 1)}$$

After thermal treatment, phosphorus atoms are titrated as orthophosphate at 460 nm by absorption spectrophotometry before (PO_4^{3-} only) and after (PO_4^{3-} and HPO_4^{2-}) acid hydrolysis of the P-O-P bond of pyrophosphate ions at 100°C for 1 hour. Thus, the level of condensed phosphate (therefore that of HPO_4^{2-}) is calculated from the difference of the results of these two analyses. The pyrophosphate content corresponding to the concentration of HPO_4^{2-} is determined within 0.5 %.

It shall be noted that this analysis method cannot be used to accurately quantify the HPO_4^{2-} content in bone and in synthetic carbonated apatites due to the presence of carbonate ions that can interfere with pyrophosphate ions and partially prevent $P_2O_7^{4-}$ formation according to equations 2 and 3 [Elliott 1994]:

$$2\ HPO_4^{2-} + CO_3^{2-} \rightarrow 2\ PO_4^{3-} + CO_2 + H_2O \qquad \text{(eq. 2)}$$

$$P_2O_7^{4-} + CO_3^{2-} \rightarrow 2\ PO_4^{3-} + CO_2 \qquad \text{(eq. 3)}$$

To open up a new alternative to the long and tedious analytical method of Gee and Dietz involving several steps and treatments, we recently set up a method to easily and rapidly evaluate the HPO_4^{2-} content in apatite, based on the mathematical decomposition of the ν_4PO_4 band of the Fourier Transform InfraRed (FTIR) spectrum [Combes 2001]. Figure 2 shows the good correlation obtained between HPO_4^{2-} determined by chemical analysis] and by FTIR spectroscopy using Gee et al. and Combes et al. methods, respectively. However, the curve-fitting parameters (decomposition of the ν_4PO_4 band, see on figure 5) need to be refined extensively to obtain a better correlation between these two methods, and thus apply this tool to rapid investigations on biological apatites.

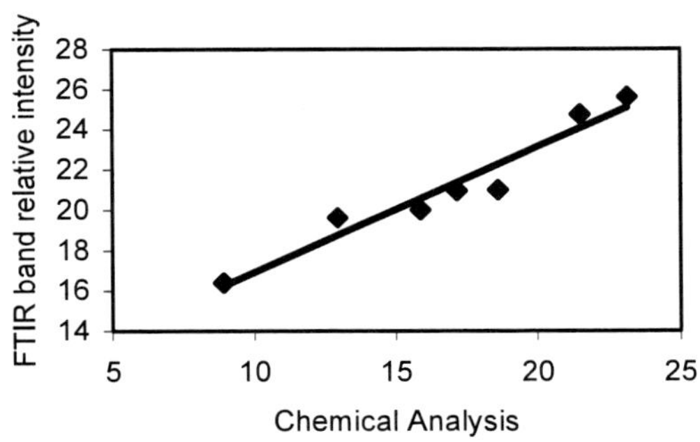

Figure 2. Correlation between HPO_4^{2-} content in apatite determined by chemical analysis and by FTIR spectroscopy [Combes 2001].

The carbonate content of apatites was determined using a CO_2 coulometer (UIC Inc., USA) that measures the CO_2 released during sample dissolution in acidic conditions ($HClO_4$,

2M) and in a closed system. The CO_2 released is transferred into a photometric cell in a non-aqueous medium and titrated through an acid-base reaction [Huffman 1977].

The determination of the amount of other ions taken up in PCA after ion exchange (strontium, magnesium ions for example) was performed using atomic absorption spectroscopy.

From the determination of the concentration of calcium, phosphate, carbonate and foreign ions if present (Mg^{2+}, Sr^{2+}), we calculated atomic ratios such as Ca/P, Ca/(P+C), or C/P in order to follow the evolution of the chemical composition of synthetic and biological PCA during maturation and/or after ion exchange processes (direct or inverse).

3.B. Diffraction Techniques

The vast majority of the diffraction studies dealing with nanocrystalline apatites and reported in the literature is based on X-ray diffraction and only few electron diffraction data are available. This could be partly explained by the strong tendency for apatite nanocrystals to agglomerate leading to broad diffuse rings [Suvorova 1999]. Also, the way that electron beams affect these rather unstable compounds has not yet been established. To our knowledge, no neutron diffraction work has been dedicated so far to such poorly crystallized compounds.

The X-ray diffraction technique applied to the study of nanocrystalline apatite specimens is often primarily used to determine their apatitic phase purity. Although the presence of secondary crystalline phases such as pyrophosphates or whitlockite can generally be distinguished from the sharpness of their characteristic XRD patterns (within the detection limit of the equipment), the detection of other phases such as octacalcium phosphate (OCP), whose diffraction pattern is rather similar to that of hydroxyapatite, or amorphous calcium phosphate (ACP), generally requires special attention. The occurrence of a sharp low-angle diffraction peak around d=18 Å can however betray the presence of OCP for concentrations above the detection limit. A background halo in the range 27-40° (λ_{Co} = 1.78892 Å), and to a lesser extent 50-60°, is evidence of the presence of ACP. A quantitative comparison with the XRD patterns obtained for mixtures of known amounts of apatite and ACP can then be used for an estimate of the ACP amount present in the specimen. Pattern fitting methods can also be used for this evaluation. This was done for example by Rogers et al. who followed the amounts of ACP and nanocrystalline apatite present in coatings formed after immersion of a titanium substrate in simulated body fluid [Rogers 2005].

Beside phase purity evaluation, X-ray diffraction studies related to nanocrystalline apatites have mostly been dedicated to the evaluation of average crystal dimensions based on line broadening analysis [Arsenault 1988, Bonar 1983, Burnell 1980, Fisher 1987]. General findings indicate that the platelet-like apatite nanocrystals exhibit an average length along the c-axis in the range 200-400 Å for a thickness of about 20-80 Å. The width of a diffraction peak is indeed dependent on the size of the crystallites constituting the sample, following a $1/\cos\theta$ mathematical law (where θ is the diffraction angle) such as the Scherrer formula [Scherrer, 1918]:

$$L_{hkl} = \frac{K\lambda}{\beta_{size}\cos\theta_{hkl}} \quad \text{(eq. 4)}$$

where L_{hkl} is the average crystallite size perpendicular to the plane (hkl), λ is the X-ray wavelength, K is a constant close to unity dependent on the particle shape, and β_{size} is the line broadening due to the size effect and θ_{hkl} is the diffraction angle corresponding to the (hkl) plane.

However, two other factors also contribute to the overall line broadening: the existence of strain within the sample, giving a broadening effect following a tgθ law, and the instrument-related broadening effect. The latter can generally be evaluated from the XRD pattern of a well-crystallized reference sample such as stoichiometric hydroxyapatite (HA), for which size and strain broadening effects are considered negligible. The line broadening due to the sample itself (β_{sample}) can then be reached from the peak width observed after elimination of this instrumental contribution (β_{instr}). However, this process depends on the geometrical shape of the peak. Gaussian, Lorentzian (Cauchy), or a convolution of the two, are mathematical functions generally used for fitting X-ray diffraction patterns. For example, if the assumption is made that diffraction peaks can be satisfactorily fitted to a Gaussian function, then the line broadening due to the sample itself is given by equation 5:

$$\beta_{sample} = \beta_{size} + \beta_{strain} = \sqrt{\beta^2_{obs} - \beta^2_{instr}} \quad \text{(eq. 5)}$$

where β_{obs} is the overall peak width observed. In this context, calculating average crystallite sizes from the sole application of the Scherrer formula requires the assumption that strain effects can be neglected, and therefore only leads to approximate values. This was emphasized in particular by Danilchenko et al. who investigated lattice strain and crystallite size in the direction of preferred orientations along the c-axis of the hexagonal unit cell in mature and well-mineralized cortical bone (femur) from a large animal (cow) [Danilchenko 2002]. These authors underlined the importance of considering lattice strain in peak profile analyses for a reliable estimation of the average crystallite size and proposed a method based on a threefold convolution of X-ray diffraction lines. They also proposed that crystallite size and strain parameters be considered as criteria for evaluating substructure variability among bioapatites. In the case of carbonated apatites synthesized at different temperatures, Baig et al. extracted the corresponding crystallite size and microstrain parameters from Rietveld analysis of XRD data, and correlated them to the evolution of the apatite metastable equilibrium solubility [Baig 1996, Baig 1999]. These authors concluded that microstrain rather than crystallite size was the dominant factor governing the solubility of such phases. The term microstrain may well in fact hide strong heterogeneities of apatite crystals composition. This phenomenon is not generally taken into account, however it is true for bone and can be related to the remodeling process. The heterogeneity in the composition of crystals induces a variation of their unit-cell dimensions and results in broadening of diffraction peaks related to the existence of unresolved superimposed peaks which can be misinterpreted as a decrease of crystal size and/or an increase of strain. Composition heterogeneity and the existence of variable solid-solutions also seem to occur for synthetic nanocrystals. The ability of apatite to give solid solutions and their surface equilibration necessities has led to the idea that even at the level of a crystal there could be some heterogeneity.

Theoretically, XRD peak analysis should be carried out by considering the integrated width based on the entire peak profile. In practice, the full-width at half-maximum (FWHM) of the considered peak is frequently used for the determination of β_{obs}. Also, whereas analysis of the whole pattern by Rietveld-like calculations leads to more accurate information, routine XRD studies from the literature usually only focus on selected diffraction lines. Due to the platelet shape of apatite nanocrystals and their elongation along the c-axis, rough estimates of crystallite size are often based on the two diffraction lines (002) and (310), see on figure 3. While the former leads to information on the length of the platelets, the latter gives an average value of their width/thickness. It must however be noted that for very immature apatites, lines (310) and (212) from the apatite structure tend to overlap due to the strong broadening size effect, which leads to a rather difficult or imprecise determination of the (310) peak FWHM. Eichert et al. reported estimated crystallite sizes drawn from the analysis of peaks (002) and (310) in the case of some biological and synthetic apatites (see table 1), and the results confirm the nanosize of these specimens [Eichert 2001]. Another example is given by Haris Parvez et al. who followed (002) and (310) line broadening for nanocrystalline apatites involved in composites containing increasing amounts of polyaspartate [Haris Parvez 2004].

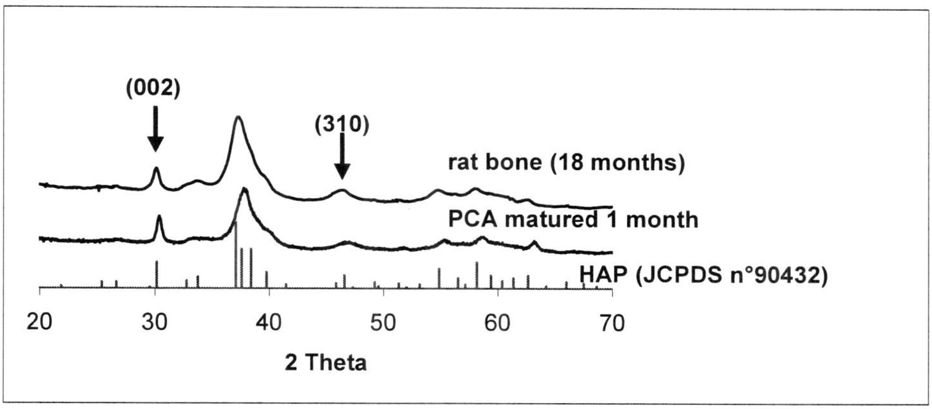

Figure 3. XRD patterns of hydroxyapatite (JCPDS reference), synthetic nanocrystalline apatite matured one month and rat bone. (λ_{Co} = 1.78892 Å).

Table 1. Estimated crystallite size for natural and synthetic apatites [Eichert 2001].

Sample	L(002) ± 3 Å (length)	L(310) ± 3 Å (width/thickness)
Chicken bone	207	66
Rabbit bone	190	Not evaluated
Adult human cortical bone	213	68
Synthetic apatite matured for 3 months	282	72

The estimation of crystal size can be advantageously used to follow the maturation state (aging in solution) of a given apatite sample. The evolution of the XRD pattern toward better resolution is indicative of the increase in crystallinity of the sample during aging in solution (data not presented). However, the pattern obtained for the apatite sample matured for 1 month still shows broad peaks, betraying the rather limited increase of crystallite size over this period (see on figure 3). Basic profile analysis applying Scherrer's formula to lines (002) and (310), and ignoring strain effects, leads to L(002) values between 145 and 280 Å, and L(310) between 50 and 70 Å, for a maturation time of apatite ranging from 0 to 3 months [Eichert 2001]. The effect of temperature on nanocrystalline apatite crystallinity and crystal size can also be advantageously observed through XRD analyses. An example is given by Shirkhanzadeh et al. who studied nanocrystalline apatite coatings prepared by electrocrystallization [Shirkhanzadeh 1994]. This author reported an increase of the crystal size from 35 nm to approximately 100 nm when heating the coatings from the synthesis temperature (65°C) to 425°C.

Nanocrystalline apatites are capable of accommodating high amounts of vacancies, leading to structural modifications of the regular apatitic lattice. The presence of structural defects can then be investigated by XRD analysis. For example, Wilson et al. performed Rietveld refinements of X-ray powder diffraction data in the case of calcium-deficient apatites, and reported the preferential loss of calcium from Ca(2) crystallographic sites rather than Ca(1) sites [Wilson 2005].

The determination of the hexagonal unit cell parameters "a" and "c" for a given specimen can give valuable information on its closeness to the regular apatitic lattice, by comparison with a reference sample like stoichiometric hydroxyapatite. XRD data are then particularly useful for following the evolution of the specimen under varying experimental conditions. For example, Panda et al. reported the increase of unit cell parameter "a" from 9.347 Å to 9.407 Å, thus approaching the reference value of 9.418 Å measured for stoichiometric HA when nanocrystalline hydroxyapatite samples initially synthesized at 80 °C by precipitation from a hydroxide gel were progressively heated up to 800°C [Panda 2003, Sudarsanan 1969]. In contrast, these authors did not observe variations of the "c" parameter in these conditions, and reported a value close to 6.862 Å to be compared to 6.881 Å for stoichiometric HA.

Variations of the lattice parameters can also be used to follow ion substitutions within the apatite structure. For example, Thian et al. recently reported in the case of a nanocrystalline hydroxyapatite coating on a titanium substrate that the substitution of Si^{4+} ions into the apatite structure resulted in an increase of both unit cell parameters "a" and "c" [Thian 2006]. These authors explained such trends by the greater ionic radius of Si^{4+} compared to P^{5+}. However, in this case, the discussion should also include composition variations related to the charge compensation mechanism. Despite this silicon substitution, no variation of XRD relative intensities was detected, although such effects are usually expected after lattice ion substitutions. This was justified here by the closeness of P and Si in the periodic table and the relatively low levels of silicon in substitution (up to 5 wt.%).

Finally, X-ray diffraction can be used for evaluating the Ca/P molar ratio of nanocrystalline apatites, and more globally of calcium-deficient apatites. The method is based on the evaluation of the amount of β-TCP formed upon heating the samples at temperatures higher than 700 °C, by comparison with XRD patterns obtained for known mixtures of β-TCP and apatite [Ishikawa 1993].

3.C. Spectroscopic Methods

The most recent and significant progresses in the characterization of apatite nanocrystals were obtained by spectroscopic methods, especially Fourier Transform InfraRed (FTIR), Raman and Solid State Nuclear Magnetic Resonance (NMR) spectroscopies.

FTIR and Raman Spectroscopies

Vibrational spectroscopies (FTIR and Raman) can deliver information on the chemical environments of phosphate, carbonate, water molecules and hydroxide ions. The theoretical vibrational modes of phosphate groups in stoichiometric apatites are shown in table 2.

Table 2. Theoretical internal vibrational modes of PO_4^{3-} ion in stoichiometric HA (R: Raman activity, IR: Infrared activity).

ν_1	ν_2	ν_3, ν_4
	Hexagonal, Space Group: $P6_3/m$	
	Site symmetry (C_s)	
1 A' (IR,R)	1A' (IR,R), 1A"(IR,R)	1A' (IR,R), 2A"(IR,R)
	Factor group theory (C_{6h})	
A_g(R), E_{2g}(R), B_u,E_{1u}(IR),	$1A_g$(R), $1E_{2g}$(R), $1B_u$, $1E_{1u}$(IR), $1B_g$, $1E_{1g}$(R), $1A_u$ (IR), $1E_{2u}$	$2A_g$(R), $2E_{2g}$(R), $2B_u$, $2E_{1u}$(IR), $1B_g$, $1E_{1g}$(R), $1A_u$ (IR), $1E_{2u}$

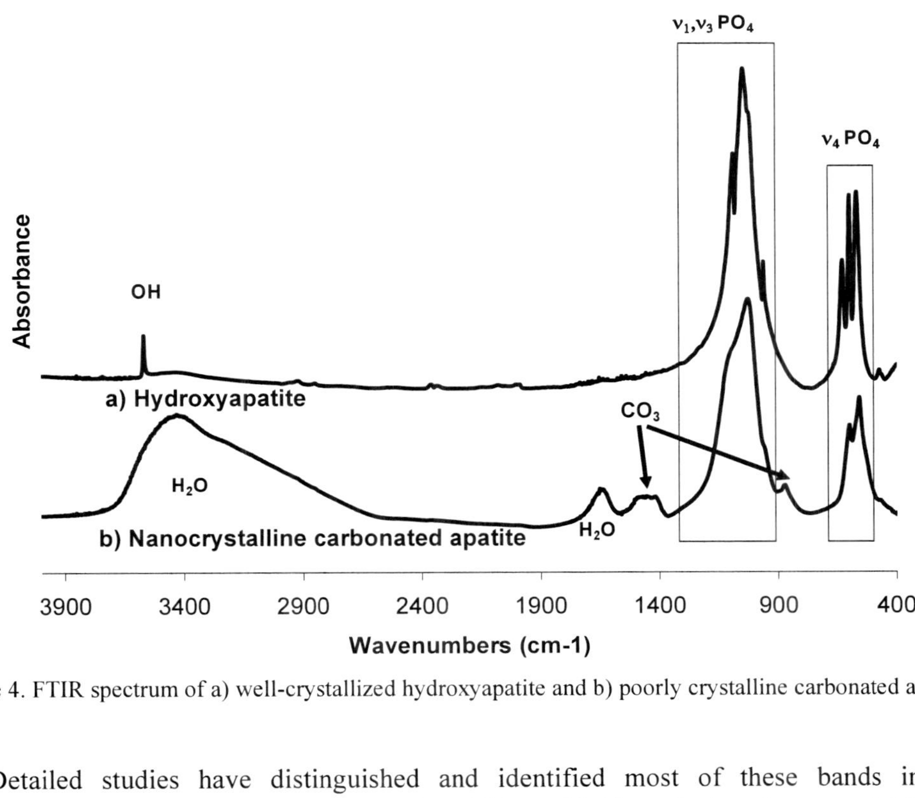

Figure 4. FTIR spectrum of a) well-crystallized hydroxyapatite and b) poorly crystalline carbonated apatite.

Detailed studies have distinguished and identified most of these bands in well-crystallized apatites [Leung 1990, Penel 1998]. In highly substituted non-stoichiometric apatites however the assignments are much less precise. Distortions of the ionic environments introduce band broadening limiting the band resolution and partly disrupt the vibrational correlation related to factor-group theory. Nanocrystalline apatite spectrum is showed in figure 4 and the phosphate band positions are reported in table 3 (comparison with stoichiometric HA).

Table 3. IR and Raman bands observed for stoichiometric HA and nanocrystalline apatite.

Domain, Assignments	Stoichiometric Hydroxyapatite		Nanocrystalline apatite	
	IR (cm^{-1})	Raman (cm^{-1})	IR (cm^{-1})	Raman (cm^{-1})
ν_2 PO_4		433		432
	464	448	469	452
	474			
HPO_4 non-apatitic			533	
HPO_4 apatitic			551	
ν_4 PO_4	567	580	562	584
	572	591	575	590
	602	607	603	611
		614		
PO_4 non-apatitic			617	
ν_L OH	633			
P—OH of HPO_4			870	873
non-apatitic			866	
ν_2 CO_3 type B			871	
type A			880	
ν_1 PO_4	964	964	962	961
ν_3 PO_4			1006	1005
			1020	
	1026	1029	1031	1032
	1034	1034		
	1044	1041	1044	1044
		1057	1059	
	1063	1064		
	1089	1077	1072	1071
			1091	
			1104	
HPO_4			1144	
ν_1 CO_3 type B				1071
ν_1 CO_3 type A				1103
B + non-ap			1420	
ν_3 CO_3 A + B			1460-1470	
non-apat			1500	
A			1540	

When HPO_4^{2-} ions are present, specific bands relative to this ion are observed.

In most nanocrystalline apatite samples and especially in biological apatites, however, additional bands are observed which do not appear in well-crystallized apatites and which have been designated as "non-apatitic environments" of the mineral ions (figure 5) [Rey 1989,

Rey 1990]. These "non-apatitic" phosphate environments have been shown to appear more clearly in the ν_4 PO_4 domain.

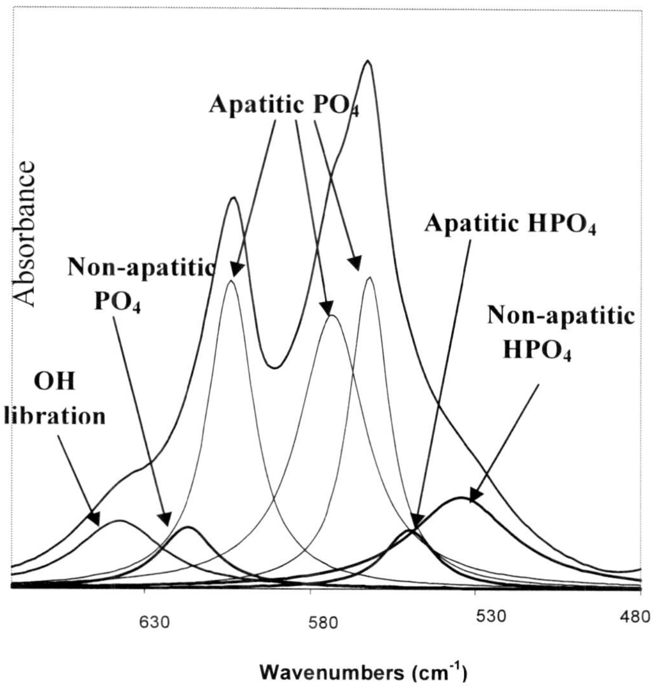

Figure 5. Example of ν_4PO_4 FTIR band decomposition (curve-fitting) using Grams/32 software (Galactic Industries Corp.).

Two distinct shoulders corresponding to measurable bands after curve-fitting have been respectively assigned to non-apatitic PO_4^{3-} groups and non-apatitic HPO_4^{2-} groups. The last assignment was confirmed using chemical analysis (see section 3.A) [Combes 2001]. The formation of non-apatitic environments has been shown to be related to the synthesis of apatite nanocrystals at physiological pH. It corresponds to exchangeable surface ions (see in the sections 4 and 5).

Carbonate ions in pure type A and type B environments exhibit specific FTIR and Raman bands (table 3). However, as in the case of phosphate groups, additional vibrational bands are found in all biological apatites and apatite nanocrystals synthesized at physiological pH, corresponding to non-apatitic environments of the carbonate ions (figure 6 and table 3). A characteristic band is clearly seen in the ν_2CO_3 IR domain and can be used for the quantitative determination of the amount of "non-apatitic" carbonate environments. The latter have been shown, using ion exchange experiments, to share the same surface domain as the "non-apatitic" HPO_4^{2-} environments (see in the sections 4 and 5).

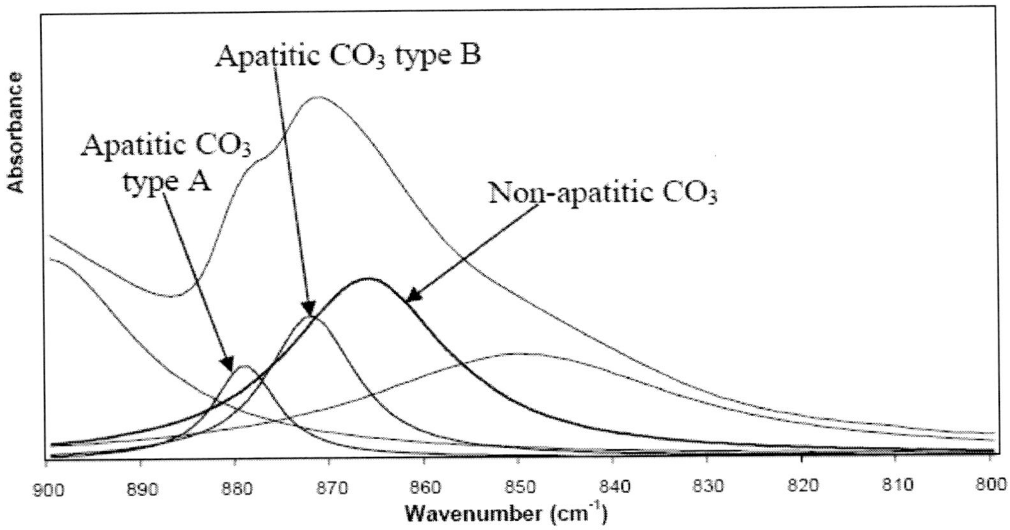

Figure 6. Example of ν_2CO_3 FTIR band decomposition (curve-fitting) using Grams/32 software (Galactic Industries Corp.).

Most characterization procedures of biomaterials have been developped for dry samples. However, biomaterials function in aqueous media, and, especially for nanocrystals, the hydrated surface might considerably differ from dry surfaces. Thus, experiments have been recently carried out on wet samples to determine possible alterations of their spectroscopic characteristics.

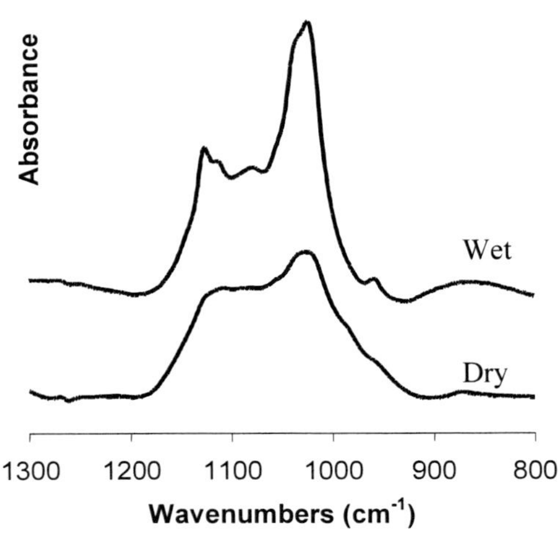

Figure 7. Effect on drying on the ionic environment as shown by FTIR spectroscopy in the $\nu_3PO_4^{3-}$ domain. Reprinted from [Rey 2006] with permission from Elsevier.

The data obtained show a considerable change (see on figure 7). The wet, freshly precipitated apatite nanocrystals exhibit very thin bands testifying to the existence of a structured hydrated layer whose characteristics seem close to those of OCP, although several differences have been noticed [Eichert 2004]. Upon drying, this fine structure is lost and considerable band broadening occurs leading to the features already described as "non-apatitic environments".

The structuration of the hydrated layer seems very sensitive to the ion content of the nanocrystal surface and it has been shown to be reversibly modified during ion exchange reactions (see section 5B). Some of the characteristics of the wet crystals spectra are summarized in table 4. These characteristic bands decrease progressively (but they never totally disappear) during the maturation of the wet crystals confirming the slow growth of apatite domains at the expense of the surface hydrated layer [Eichert 2001, Eichert 2004].

Table 4. Positions et assignments of FTIR bands for wet precipitated PCA (w: weak, vw: very weak; sh: shoulder).

Assignments	Band positions (cm^{-1})
HPO_4 (OH in plane bend)	1195 $_{w,sh}$
Stretching ν_3 HPO_4	1137 $_{w,sh}$ 1127 1110
Stretching $\nu_3 HPO_4$, ν_3 PO_4	1075 $_{w,sh}$
Stretching ν_3 PO_4 Stretching ν_3 PO_4 Stretching ν_3 PO_4	1055 $_{w,sh}$ 1035 1022
Stretching $\nu_1 HPO_4$	1000 $_{vw,sh}$
Stretching $\nu_1 PO_4$	961
Stretching HPO_4 (P-OH)	860

Solid State NMR

For over twenty years solid state NMR has been widely used to investigate the chemical structure (bulk and surface) and structural properties of calcium phosphates and related biomaterials. In particular, this technique provides detailed information on chemical and structural environments of phosphate, carbonate and hydroxyl groups, by observing single and/or double resonance of different nuclei (^1H, ^{31}P and ^{13}C) [Rothewel 1980, Tropp 1983, Aue 1984, Yesinovski 1987, Belton 1988, Miquel 1990, Beshah 1990, Pan 1995, Cho 1996, Sfihi 2002, Wilson 2005, Jarlbring 2006, Jäger, 2006]. In a stoichiometric apatite and related materials, the ^{31}P chemical shift of all the phosphate groups has been reviewed [Yesinovski 1998]. In stoichiometric apatite, all PO_4^{3-} groups have the same chemical environment and therefore only a single line is observed at 2.3-2.9 ppm with respect to 85 % H_3PO_4 [Rothewel 1980, Tropp 1983, Aue 1984, Belton 1988, Miquel 1990, Sfihi 2002, Wilson 2005, Jarlbring 2006, Jäger, 2006]. However, in some cases, in particular when the samples are not well crystallized, this line is separated into two components corresponding to PO_4^{3-} groups in crystalline and amorphous phases, by combining relaxation and double resonance experiments [Isobe 2002]. In bone nanocrystals, the study of chemical shift anisotropy suggests P environments different from those of apatites [Roufosse 1984, Roberts 1992, Wu 1994]. In addition, in apatite nanocrystals an inhomogeneous and broad ^{31}P peak of the main phosphate is generally observed [Wu 1994, Jarlbring 2006]. This peak, related to HPO_4^{2-}, can be isolated using 2-D $^{1}H\leftrightarrow^{31}P$ solid state NMR [Wu 1994].

^{31}P solid-state NMR spectra of wet nanocrystals are very different from those of dry crystals. Indeed, they exhibit different peaks located in the 0 - 6 ppm range assigned to different phosphate species [Eichert 2004]. These specificities are consistent with the FTIR observations, but the analogy with OCP, as could have been expected, is not attested. The HPO_4^{2-} peak seems the most affected suggesting that these species are essentially located in the hydrated layer (see section 4). The characteristics of the spectra suggest possible proton hopping between phosphate groups of the hydrated layer.

Because of the very low carbonate content, the studies of carbonate ions generally needed enrichment with ^{13}C for NMR observation in a reasonable time. Type A and Type B carbonated apatites show two distinct peaks located at 166 and 170 ppm with respect to tetramethylsilane (TMS), respectively [Beshah 1990]. In addition "non-apatitic" carbonate can exhibit a broad peak at *ca.* 168 ppm which has been shown to be associated with water molecules [Sfihi 2002].

^{1}H solid state NMR measurements have been performed especially to detect hydroxide ions. In hydroxyapatite, the OH^- groups give a single ^{1}H NMR peak at approximately 0.2 ppm with respect to TMS [Yesinowski 1987]. This peak has been detected at the same position in bone [Cho 2003]. However, the indirect ^{1}H solid state NMR measurements used to detect the OH groups in bones did not allow an evaluation of their amount. Other ^{1}H peaks are related to water molecules again associated with bone apatite nanocrystals [Cho 2003]. Similar results were recently reported in synthetic nanocrystalline apatites [Cho 2003]. This peak, which appears in most calcium phosphate bioamaterials and which is centered at *ca.* 6 ppm, is relatively broad and probably corresponds to several labile environments. It seems dominated by water molecules from the surface layer (see the model presented section 4). Additional unresolved ^{1}H broad peaks of weak intensities assigned to HPO_4^{2-} ions can also be observed in the range 10 – 16 ppm both in bone mineral [Cho 2003] and in synthetic nanocrystalline apatites [Jäger 2006]. These peaks testify to the existence of several HPO_4^{2-} environments, but their study needs to be completed, to specify in particular their localisation.

X-Ray Absorption Spectroscopies (XANES and EXAFS)

As illustrated above, the environments of anions in nanocrystalline apatites can be characterized using different spectroscopic techniques. In contrast, the environments of calcium ions can only be observed through techniques involving core electron transitions which are rather difficult to carry out. Several sets of data have however confirmed the existence in non-stoichiometric nanocrystalline apatites of specific cationic environments that are absent in well crystallized stoichiometric apatites and assigned to "non-apatitic" environments of calcium ions [Eichert 2005]. The existence of these environments can be related to cation exchange properties of apatite nanocrystals.

4. A MODEL FOR NANOCRYSTALLINE APATITES

Nanocrystalline apatites, whether biological or synthetic, exhibit an extended surface area due to the nanosize of their constitutive crystals. As for any other kind of nanomaterials, the surface-to-volume ratio is therefore high and all experimental results must then be considered as a combination of bulk and surface contributions. The importance of surface behavior is even more crucial for biomedical applications since numerous functions of the bone mineral involve at the interface between the surface of such apatite nanocrystals and the surrounding biological fluids.

In addition to the use of characterization techniques specifically adapted to surface investigations, experimental results obtained with non-surface-specific techniques can also be exploited and were indeed found to be of prime importance for the exploration of the nanocrystalline apatites surface structure. Section 3 reported results of our investigations using spectroscopic techniques (FTIR and solid state NMR) revealing the presence of non-apatitic chemical environments in the case of synthetic as well as biological nanocrystalline apatites.

We can summarize that according to the results of complementary investigations on nanocrystalline apatites using spectroscopic techniques and chemical analysis methods presented in the previous section: (1) apatite nanocrystals involve non-apatitic anionic and cationic chemical environments, (2) at least part of these environments (e.g. labile carbonates) are located on the surface of the nanocrystals and are in strong interaction with hydrated domains, (3) immature samples show IR band fine substructure that is altered upon drying without leading to long-range order modifications, and (4) this fine substructure shows striking similarities with the FTIR signature of OCP which is constituted by alternating "apatitic" and "hydrated" layers.

All these elements favor a model in which apatite nanocrystals are covered with a rather fragile but structured surface hydrated layer containing relatively mobile ions (mainly bivalent anions and cations: Ca^{2+}, HPO_4^{2-}, CO_3^{2-}) in "non-apatitic" sites. However the exact structure and composition of this hydrated layer are still under investigation. The existence of a hydrated layer with a composition different from that of bulk apatite domains can modify the description of apatite composition presented in section 1. A schematic representation of this "surface hydrated layer model" is given on figure 8.

It is interesting to remark that the two calcium phosphate phases with which spectroscopic similarities were found (DCPD and OCP) are hydrated phases and exhibit the lowest surface tensions among the examples of calcium orthophosphates studied in the literature. One could thus postulate that the formation of a hydrated layer on the surface of apatite nanocrystals could contribute to the reduction of the surface energy of the nanocrystals.

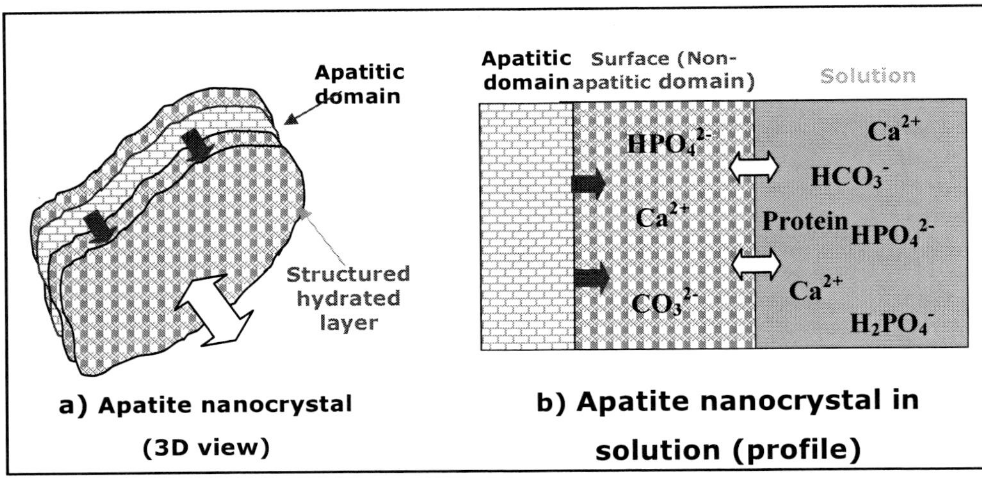

Figure 8. Schematic representation of the " surface hydrated layer model" for poorly crystalline apatite nanocrystals.

This layer might be close although not identical to that existing in octacalcium phosphate [Eichert 2004]. It differs from OCP by its ability to accomodate carbonate ions and other mineral ions like Mg^{2+}. It is thought to be responsible for most of the properties of apatites, especially high electrical conductivity probably resulting from a high superficial ionic mobility. The consideration of this type of surface state can help to understand and explain the behavior of biological apatites in participating in homeostasis due to the very high specific surface area of bone crystals and also in constituting an important ion reservoir with an availability that depends on the maturation state. The consequences of this particular mineral structure have not been completely explored, it is however plausible that bone mineral is also involved in some hormonal regulations or in chemical interactions with the organic matrix that could determine its mechanical properties. Nanocrystalline apatite properties are presented in detail in the next section.

5. Physico-Chemical Properties of Nanocrystalline Apatites

5.A. Maturation

Poorly crystallized, non-stoichiometric apatites are not thermodynamically stable and show a clear tendency to evolve in solution toward stoichiometry and greater crystallinity [Cazalbou 2004]. As indicated by the arrows in figure 8a, upon aging in solution the bulk apatitic domains slowly extend at the expense of the hydrated domains at the surface of the nanocrystals and can irreversibly incorporate some of the ions of the hydrated layer. This evolution is referred to as "maturation" or "aging in solution". It can be followed by several physico-chemical measurements.

The overall chemical composition of the initial phase varies with maturation time. For example, the changes in composition observed for a carbonated nanocrystalline apatite as a function of maturation time in the synthesis solution at room temperature (RT) are reported in table 5.

Table 5. Evolution of the chemical composition (atomic ratios) of a carbonated nanocrystalline apatite with maturation time (in synthesis solution, at RT).

Maturation time	Ca/P (± 0.02)	Ca/(P+C) (± 0.02)	C/P (± 0.02)
0	1.44	1.41	0.030
6 hours	1.51	1.42	0.060
1 day	1.51	1.40	0.089
3 days	1.54	1.39	0.145
10 days	1.62	1.40	0.162
4 months	1.62	1.38	0.200

In this case, the increase of the Ca/P molar ratio from 1.44 to 1.62 during the first steps of maturation is accompanied by the constancy of the Ca/(P+C) ratio and the increase of the C/P ratio. These trends indicate that over this time period the total number of vacancies remains unchanged for carbonated apatites while HPO_4^{2-} ions get replaced by CO_3^{2-} ions. These conclusions are confirmed by FTIR spectroscopy analysis (figure 9a and 9b), showing the progressive decrease of the non-apatitic HPO_4^{2-} content (and labile carbonates) and the concomitant increase of apatitic carbonate species (mostly B-type).

Maturation is a physico-chemical process, initiated by a thermodynamical driving force, and leading to modifications of the phase composition and structure. As such, it should be distinguished from Ostwald's aging. The initial nanocrystalline phase, rich in (surface) non-apatitic environments and poor in carbonate evolves toward a more stable structure with growing apatitic domains, in contrast with the surface hydrated layer which tends to disappear. The structure evolves toward higher stability and therefore lower solubility. The chemical composition of synthetic apatites is thus controlled by the synthesis conditions.

In parallel to chemical composition changes upon maturation, other physico-chemical characteristics also evolve. In particular, an increase in crystallite size can be clearly seen (figure 10).

In this context, the maturation state of nanocrystalline apatites has to be considered as a major parameter in the preparation of nanocrystalline apatite-based biomaterials, as it controls important factors after implantation such as dissolution rate. The surface reactivity (see next section) is also modified by the apatite maturation.

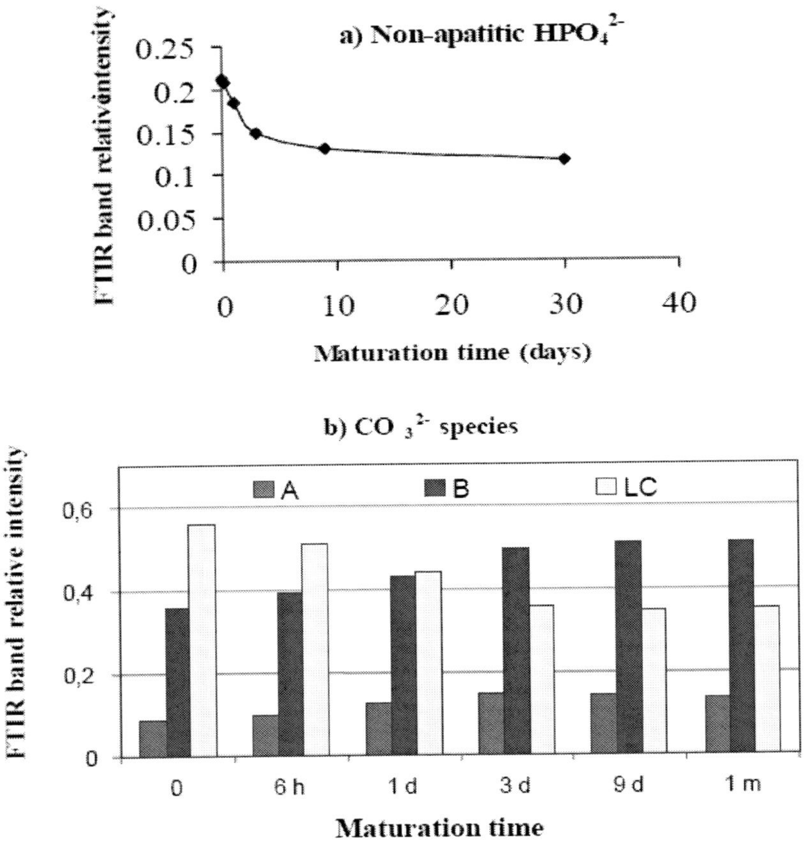

Figure 9. Evolution of a) labile HPO_4^{2-} and b) CO_3^{2-} (A: A-type carbonate, B: B-type carbonate, LC: labile carbonate species) contents with maturation time for a carbonated nanocrystalline apatite, as determined by FTIR spectroscopy.

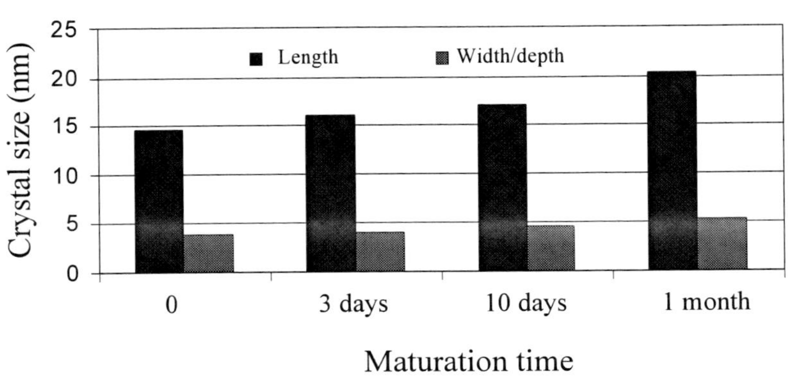

Figure 10. Variation of nanocrystalline apatite crystallite size with maturation time.

In parallel to chemical composition changes upon maturation, other physico-chemical characteristics also evolve. In particular, an increase in crystallite size can be clearly seen (figure 10).

In this context, the maturation state of nanocrystalline apatites has to be considered as a major parameter in the preparation of nanocrystalline apatite-based biomaterials, as it controls important factors after implantation such as dissolution rate. The surface reactivity (see next section) is also modified by the apatite maturation.

5.B. Ion Exchanges on Nanocrystalline Apatites

Ion exchanges in bone mineral have been the object of numerous studies [Neuman 1956, Pak 1967, Neuman 1968, Johnson 1970, Fernandez Gavarron 1978, Neuman 1985]. Indeed, these processes have been suggested to be involved in the regulation of mineral ion concentrations in body fluids (homeostasis). In this view, the bone mineral can be considered as a "reservoir" for mineral ions, capable of entrapping or releasing them as required for concentration regulation. The above-cited studies concerned both cationic (calcium) and anionic (phosphate and carbonate) exchanges. Unsurprisingly, such ion exchange processes originate at the interface between apatite nanocrystals and the surrounding biological fluids.

The first studies of ion exchange phenomena on bone apatite were mostly dedicated to phosphate/carbonate substitutions, and these investigations were first initiated in view of explaining the presence of carbonate ions in bone. Since then, the incorporation during the mineralization process, of CO_3^{2-} ions into the apatitic lattice by substitution of either phosphate (B-type) or hydroxyl (A-type) ions has been identified and studied in detail (see section 3c).

Non-apatitic carbonate environments, also referred to as "labile carbonates", have been identified in immature samples (see figure 6 and table 3). Recent works have pointed out the possibility for nanocrystalline apatites to exchange such labile CO_3^{2-} ions with phosphate groups through ion exchange in a solution enriched in a phosphate salt. Hina et al. have investigated such exchanges, and found that soaking a nanocrystalline apatite sample in a carbonate-rich solution led to an increase of the carbonate content of the solid accompanied by a decrease of its non-apatitic HPO_4^{2-} content [Hina 1996]. Similarly, when the precipitate is immersed in a phosphate-rich solution, the amount of non-apatitic HPO_4^{2-} contained in the solid increased at the expense of the carbonate content. Interestingly, such non-apatitic carbonate/non-apatitic HPO_4^{2-} exchanges were found to be highly reversible (figure 11). This is an additional point in favor of surface exchange phenomena involving easily accessible sites, most probably within the nanocrystals superficial hydrated layer. It is important to note that the apatitic carbonate species are not altered by subsequent immersion in phosphate solution. This was for example shown by Hina et al. using $^{13}CO_3^{2-}$ ions [Hina 1996]. This emphasizes the distinction between the ions that are located in the surface layer and are therefore rather mobile, and those located in stable apatitic sites in the bulk (see figure 8 and section 4).

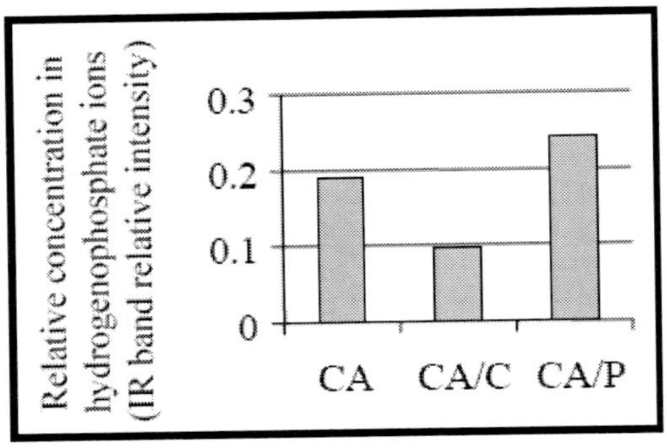

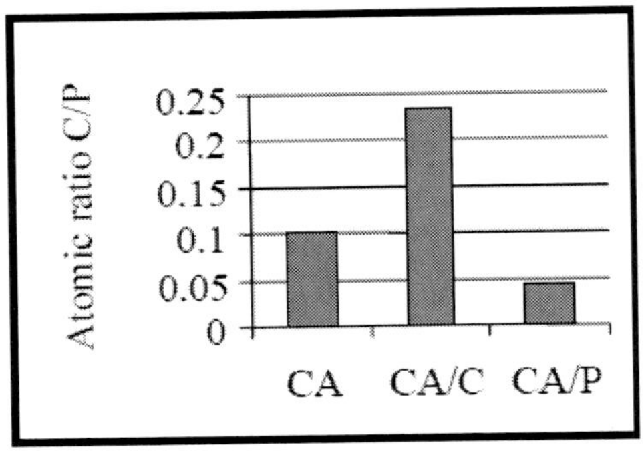

a) **b)**

Figure 11. Example of ionic exchange (carbonate-phosphate) experiments: a) Evolution of hydrogenphosphate ion concentration (IR data) b) Evolution of carbonate ion concentration (C/P ratio calculated from chemical analysis results). CA: initial carbonated apatite matured 24h, CA/C: CA immersed in ammonium carbonate solution (1M, 10 mn); CA/P : CA immersed in ammonium phosphate solution (1M, 10 mn) [Cazalbou 2004] - Reproduced by permission of The Royal Society of Chemistry.

The capacity to undergo such surface ion exchanges is however highly dependent on the maturation state of the apatite specimen. Indeed, the total amount of exchangeable carbonate ions (i.e. labile carbonates) was found to decrease with maturation time as indicated in figure 12. This phenomenon is attributed to the progressive disappearance, upon maturation, of the hydrated surface layer of nanocrystalline apatites which is thought to accommodate the labile carbonate species.

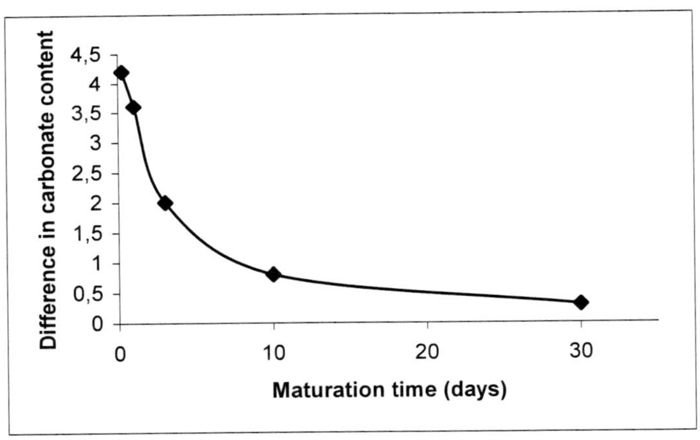

Figure 12. Decrease in exchangeability of carbonate ions versus maturation time.

Beside anionic exchanges, illustrated above by the carbonate/phosphate pair, cationic exchanges (substitution of calcium) also deserve special attention. In this regard, ions such as Mg^{2+} and Sr^{2+} are of particular interest due to their biological activity. Magnesium is one of the most abundant foreign elements in biological hard tissues and is thought to play a role on bone cell adhesion [Tsuboi 1994]. Strontium, although a minor constituent of bone, was shown to play an important role in osteoclast cell activity and could therefore be used in the treatment of some bone pathologies [Pors 2004]. Marie et al. showed for example that strontium had pharmacological effects, including antiresorptive and anabolic activity, leading to possible interest for the treatment of osteoporosis and other osteopenic pathologies [Marie 2001].

The works reported by Cazalbou et al. and Eichert et al. show that magnesium and strontium ions can be entrapped by nanocrystalline apatites by immersion in a solution containing the exchanging ion [Cazalbou 2000, Eichert 2001]. The amounts of Mg^{2+} and Sr^{2+} taken up were found to decrease with apatite maturation (figure 13). The reversibility of these exchanges was also investigated. Most of the magnesium taken up after ion exchange in a Mg-rich solution was fixed in a reversible way whatever the maturation stage of the apatite (see on figure 13a). In contrast, the amount of reversibly-fixed Sr^{2+} ions decreased noticeably with maturation in a Sr-containing solution (see on figure 13b). These results suggest that Mg^{2+} ions are mostly entrapped on surface sites whereas Sr^{2+} ions are progressively irreversibly incorporated (beyond 24 hours of maturation) into the apatitic domains of the maturating nanocrystals.

Recently the cationic exchange ability of non-carbonated and a carbonated nanocrystalline apatite, matured for one day, for magnesium and strontium ions was studied in depth. The Langmuir-like isotherms for magnesium ions are reported in figure 14, and indicate that the maximum amount of Mg exchanged is greater for the carbonated sample. These results were attributed to a greater proportion of hydrated layer on the carbonated apatite sample, due to the well-known apatite crystal growth inhibiting effect of carbonate ions. Similar results are obtained for strontium ions but Sr uptake is noticeably greater than that of Mg for both kinds of apatites (data not presented). The difference in behavior of Mg and Sr ions is still under consideration. Possible reasons include the occupancy of different cationic surface sites and differences in stability of the complexes $[Me(H_2O)_6]^{2+}$ (Me = Mg, Sr) in solution due to the larger size of Sr^{2+} ions.

The reverse exchanges were also investigated by soaking the pre-exchanged samples in a calcium-rich solution. In these conditions, ca. 85% of the Mg ions and 75-80% of Sr ions were released upon inverse exchange, indicating that on apatites matured one day most of the magnesium and strontium remained exchangeable, and therefore most probably located on surface sites within the nanocrystal hydrated layer rather than in apatitic domains of the bulk. It is also important to note that no secondary phase was observed by XRD analysis after such ion exchange experiments.

As a concluding remark on ion exchanges on nanocrystalline apatites, it is of interest to distinguish the two categories of exchange ions. A first type of ion can be incorporated into the surface hydrated layer of the nanocrystals and can then progressively replace apatitic ions in the bulk. This is the case of strontium. The second type of ion cannot massively substitute for bulk ions and they mostly remain located at the nanocrystal surface, like magnesium. These considerations, once generalized, have several implications especially for biological nanocrystalline apatites. As apatite maturation is an inevitable physical-chemical phenomenon leading to the loss of vital ion exchange properties, it is essential that bone mineral is renewed [Rey 1995]. However with aging, the bone remodeling cycle becomes slower, and therefore superficial properties are predictably degraded. This evolution explains

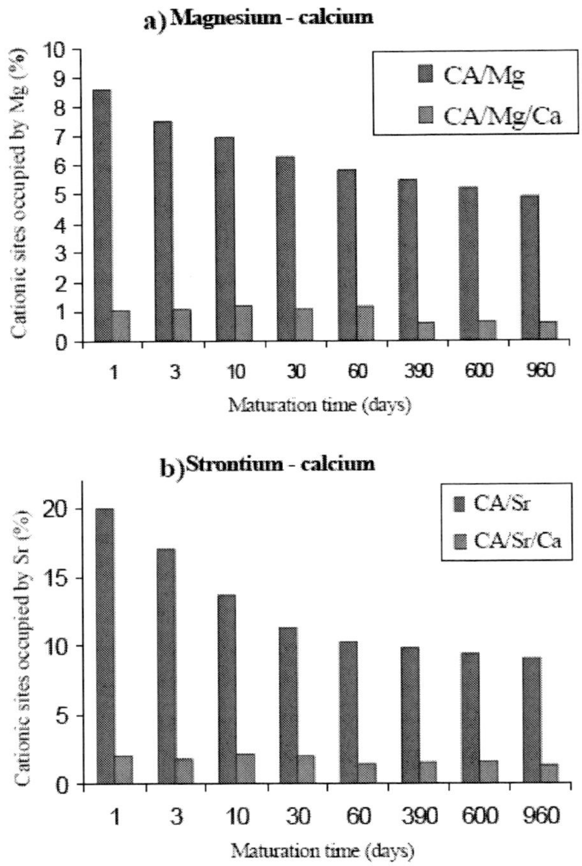

Figure 13. a) Mg and b) Sr uptake versus nanocrystalline apatite maturation time.

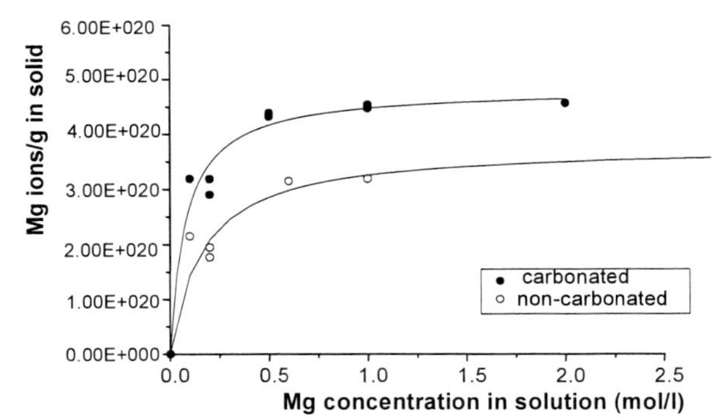

Figure 14. Magnesium uptake on 1 day-maturated nanocrystalline apatites (carbonated or non-carbonated) after direct exchange with Ca^{2+}.

the higher sensitivity of children to poisoning by some mineral ions and the preferential location of these contaminants in areas of fast bone remodeling (epiphysis) [Cazalbou 1999]. For example, intoxication with heavy metals is a well-known health issue especially problematic for young children and among such elements is lead (Pb^{2+} ions exhibiting an ion radius close to that of strontium). It is therefore likely that these ions could reach the surface of apatite nanocrystals through blood transport, and be entrapped in the surface sites. The higher sensitivity of children to lead intoxication could then be due to their less matured bone mineral, offering greater proportions of non-apatitic environments available for rapid ion exchanges. This stresses the importance of rapid intervention after such lead poisoning, in order to proceed with inverse exchange with calcium before the Pb^{2+} ions reach stable apatitic substitution sites. Indeed, nearly 50% of any strontium uptake is inaccessible to further ion exchange after maturation of 24 hours [Cazalbou 2000].

5.C. Protein Adsorption on Nanocrystalline Apatites

A good knowledge of the bone mineral physico-chemical characteristics is essential to understand its biological functions as well as to develop engineered biomaterials aimed at bone repair. However, this needs to be completed by the analysis of surface reactivity toward biological fluids. The previous section concerned mineral ion exchanges involving the surface hydrated layer of apatite nanocrystals. Other major components of body fluids are proteins. Since protein adsorption is known to precede cell adhesion to bone apatite, such adsorption processes also appear as critical steps and deserve special attention.

The first interest in protein adsorption arose from the observation that mineralized matrix proteins could initiate mineralization and regulate crystal growth. Several studies dealt with non-collagenous proteins such as albumin or phosphoproteins [Termine 1980, Glimcher 1990]. Attempts to classify these proteins as biomineralization promoters or inhibitors failed as other parameters like protein concentration and conformation were found to play a key role in such processes. Combes et al. reached similar conclusions during the study of the effect of bovine serum albumin (BSA) on the nucleation and growth of octacalcium phosphate (OCP) on type-I collagen. In this case, BSA was found to play alternately a promoter effect at low concentrations and an inhibitor effect at high concentrations ($> 10\ g.L^{-1}$) [Combes 2002].

In contrast to the large number of studies reported on protein adsorption on well-crystallized hydroxyapatite or other calcium phosphates, only few data are available for nanocrystalline apatites that constitute the most abundant mineral component of bone tissue. Ouizat et al. investigated the effects of BSA adsorption on such poorly crystallized apatites [Ouizat 1999]. In this work, the authors concluded that non-apatitic environments, in particular labile phosphate groups, were involved in the protein binding process, and found that the adsorption capacity decreased upon apatite maturation (table 6). On the other hand, a higher affinity for BSA was observed in matured samples, which was related to possible competition between water and BSA molecules for the interaction sites on the apatite crystals, and to the presence of non-apatitic phosphate groups which enhance electrostatic repulsion at the surface.

Table 6. Influence of maturation time on the adsorption of BSA onto nanocrystalline apatites in 1 mM KCl solution at pH ≅ 7.

Apatite samples	Maximum amount adsorbed ($mmol/m^2$)	Affinity constant (mL/mg)
Not matured	42.1 ± 5.3	0.5 ± 0.2
Matured 15 days	33.6 ± 1.2	39.0 ± 17.2
Matured 45 days	23.4 ± 0.7	229.9 ± 81.8

The protein adsorption ability of apatites can be used in the biomaterials field to improve the bone wound-healing processes by associating apatite (carrier) with growth factors which have been shown to be potent osteogenic inductors or with a specific protein RGD sequence (arginine-glycine-aspartic acid) involved in cell attachment to favor cell adhesion, proliferation and differentiation. Midy et al. reported a study on the adsorption and release properties of synthetic carbonated apatite (matured for 24h) analogous to bone mineral compared to the properties of well-crystallized hydroxyapatite toward basic fibroblast growth factor (bFGF) [Midy 1998]. The adsorption of bFGF was much greater on carbonated apatite (85% of the amount contained in solution) than on hydroxyapatite (10% of the amount contained in solution) probably due to the higher amount of non-apatitic environments of carbonate and hydrogenphosphate ions in carbonated apatite. Indeed it has been suggested that non-apatitic environments and proteins share the same adsorption sites at the surface of the apatite nanocrystals [Rey 1994]. However, the proportion of bFGF released remained low and quite the same for the two kinds of apatite (from 17 to 27 % of the adsorbed amount) probably due to strong interactions between bFGF and apatite surfaces through acidic residues of this growth factor (negatively charged) and also to the short duration of the experiment (60 min) related to the short growth factor half-life. This study showed that growth factor binding to apatites depends strongly on apatite characteristics (poorly crystallized non-stoichiometric apatites rich in non-apatitic surface species are more efficient than stoichiometric well-crystallized HA). In vivo, these properties could play a role in signaling the evolution of the mineral crystals (apatite) to cells and in bone remodeling.

It is interesting to note that interactions between biomimetic apatites and proteins can, in some conditions, lead to physico-chemical alterations of the initial solid phase. In this view, Luong et al. have recently investigated the effect of BSA on a biomimetically nucleated mineral, and showed that while surface adsorption of BSA did not lead to structural modifications of the mineral, BSA incorporation (occurring when added during the synthesis) changed the crystal morphology from plate-like to more rounded structures: similar observations have been made by Combes et al. in the case of OCP nucleation and crystal growth in the presence of BSA [Combes 2002, Luong 2006].

To gain a better understanding of the adsorption mechanisms on apatite nanocrystals, smaller molecules such as peptides or amino acids are often used for closer identification of possible interacting groups. It is thus generally accepted from experimental results that the alpha-carboxylate group of the amino acid is preferentially bound to the nanocrystal surface. The affinity was also found to be dependent on the apatite characteristics. For example in the case of glycine, the affinity decreased as the HPO_4^{2-} content decreased and as the carbonate content increased [Bennani-Ziatni 2003]. Also, the affinity was found to increase with the proportion of ethanol in the medium, i.e. with the decrease in dielectric constant. The effect of the presence of phosphate ions in the medium was investigated for alanine and phenylalanine adsorption [Bihi 2002]. It was then found that the phosphate ions hindered the adsorption process by competing with the amino acid carboxylate group for interacting with the calcium ions at the nanocrystal surface. Another interesting result arose from the comparison of adsorption parameters for serine and phosphoserine [Benaziz 2001]. In the latter case, a greater affinity was found, which was related to the presence of a phosphate group in this amino acid molecule, that has specific attachment sites. Evidence of the involvement of anionic groups of the amino acid in the adsorption process is in particular given by a shift of the related IR bands, stressing the close interaction with surface species on the crystals (e.g. shift for the carboxyl groups). Other functional groups of amino acids can however also be involved in the adsorption process. For example, the amine groups of glycine were also found to intervene [Bennani-Ziatni 1997].

Finally, the nature of the amino acid was found to play a role in the amount that can be bound on biomimetic apatites. Indeed, Bihi et al. studied the adsorption of alanine and phenylalanine on such compounds, and observed that the increase of the size of the amino acid side chain led to a decrease of maximum uptake at saturation [Bihi 2002].

6. Processing of Nanocrystalline Apatite-Based Biomaterials

Nanocrystalline calcium phosphate apatites are interesting biomaterials due to their surface reactivity which has just started to be explored. The existence of a superficial hydrated layer including highly labile ions offers an opportunity to vary and adapt the surface properties, whether for materials science or for biological purposes. Thus, it is possible to take advantage of the reactivity of nanocrystalline calcium phosphate apatites in the processing of bioactive bone substitutes.

As illustrated in figure 15, one of the most interesting properties related to the existence of the hydrated surface layer is the ability of nanocrystals to join and strongly interact with each other and/or with different substrates and macromolecules. Interestingly, even though it is not yet fully understood, the junction of apatite crystals has been reported in living organisms for example during dental enamel formation and is referred to as "crystal fusion" [Daculsi 1984].

In the materials science field, the hydrated layer can be involved in cohesiveness and adhesion between two apatite nanocrystals or between apatite nanocrystals and different substrates in the low temperature processing of ceramics and coatings or composites respectively. The progressive drying increases inter-crystal or crystal-substrate contacts. Upon drying, the steady elimination of excess water molecules brings two crystals together enabling the constitutive ions to interact from a strong electrostatic interaction. At the end of the process, crystals that have been joined cannot be split apart by simple rehydration. Intrinsic and extrinsic parameters such as the extent of the hydrated layer at the surface of the nanocrystals (depending on the apatite maturation state) and the rate and temperature of drying, respectively, have to be considered to optimize the processing of nanocrystalline apatite-based biomaterials.

Depending on the type of repair, surgeons need bioceramics processed in different forms (dense or porous blocks, coatings, cements) in which apatites can be used alone, or as part of a composite, or formed as a result of a reaction in situ (bone cements). We present hereafter a short review of nanocrystalline apatite-based cements, coatings and composites biomaterials.

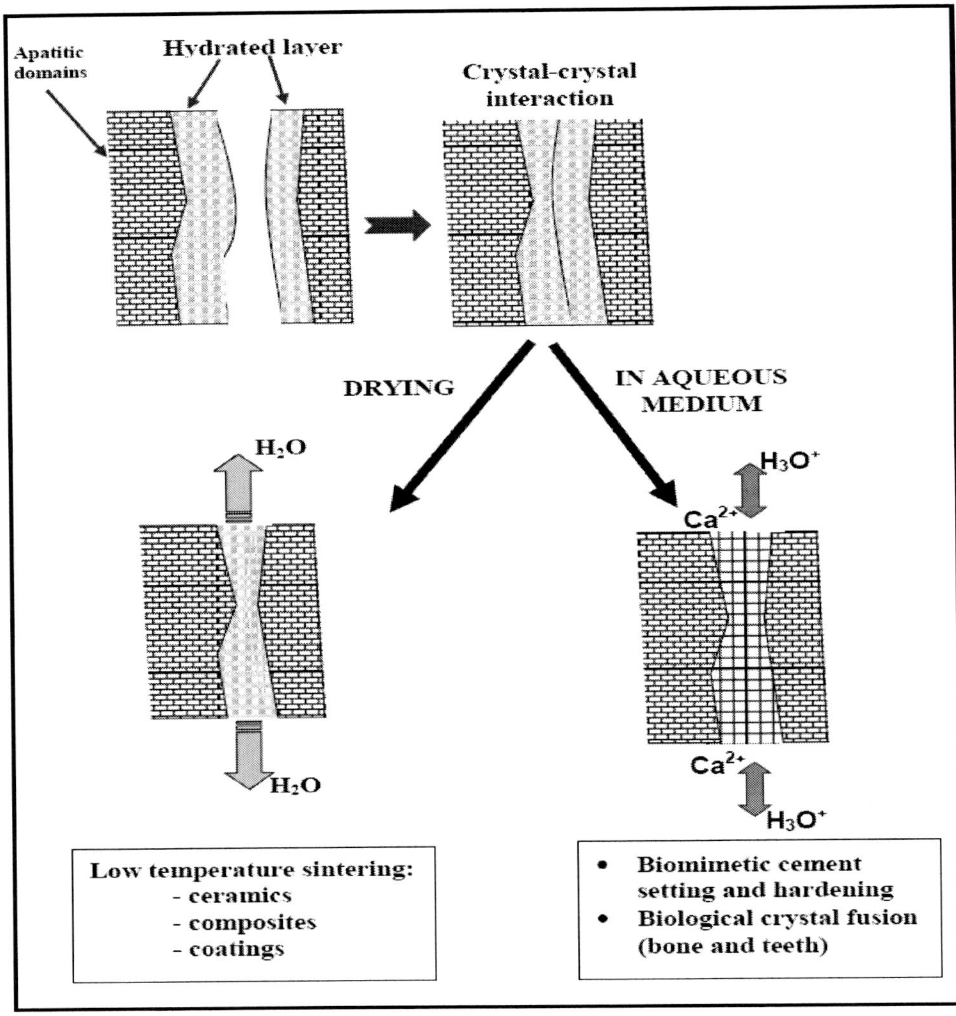

Figure 15. Role of the hydrated layer in the preparation of materials involving poorly crystalline apatite nanocrystals.

6.A. Biomimetic Cements

Biomimetic calcium phosphate (CaP) cements are based on the ability of CaP phases to form apatitic calcium phosphates in aqueous media. Different types of cements may be distinguished depending on the kind of reaction involved (table 7).

Acid-Base Cements

Most cements rely on an acid-base reaction between calcium phosphate phases. The acidic phase can be either a soluble CaP salt such as monocalcium phosphate or even phosphoric acid, or an insoluble CaP salt at physiological pH like dicalcium phosphate dihydrate (DCPD). The alkaline phase is generally tetracalcium phosphate (TTCP) but other phases like calcium carbonate or even calcium oxide have also been proposed.

The setting reaction in these cements is often complex and difficult to control as each of the constituents can hydrolyse separately in aqueous media into apatite. The pH of the paste can undergo strong variations during the setting time from acidic to alkaline. The particle size is a critical parameter which determines the setting characteristics of the cement. Generally the setting reaction involves several intermediates such as DCPD and/or OCP [Hatim 1998]. A multitude of formulations may be proposed for these cements and they offer a wide range of possibilities and characteristics. From a thermodynamic point of view, the acid-base reaction is always exothermic although the release of heat is generally much less than that of the well-known poly-methylmethacrylate (PMMA) biomedical cements [Baroud 2006].

Monocomponent Cements

Monocomponent CaP cements are based on the fast hydrolysis of one calcium phosphate salt into apatite. In order to achieve the short setting times needed for orthopaedic applications, the hydrolysis reactions have to be fast and the CaP phases should be rather unstable. Amorphous CaP and alpha-tricalcium phosphate (α-TCP) have been proposed [Lee 2000, Ginebra 1997]. The monocomponent cements induce less pH variation during setting and there is a direct conversion into an apatitic phase.

Generally the rate of the hydrolysis reaction is determined by the temperature. For amorphous calcium phosphate, for example, an increase of temperature causes an increase in the rate of conversion into apatite nanocrystals thus, at 20°C the paste does not harden and it is only at body temperature that fast hardening is achieved. These cements can easily be injected [Knaack 1998].

A third class of CaP cements has been developed that consists of unstable phases at physiological pH that transform into apatite after implantation (see table 7, hydrolyzable phase-based cements). The most frequently encountered examples are based on dicalcium phosphate dihydrate (DCPD) [Mirtchi 1989]. These are acid-base cements and the setting reaction corresponds to the formation of DCPD, which possesses a crystallographic structure analogous to that of gypsum or plaster of Paris. DCPD is an unstable CaP phase at neutral pH, and transforms into apatite at physiological pH.

Sometimes both the first (acid-base) and second type (conversion of an unstable CaP phase) of setting reactions are involved. Additives may often be used in all classes of CaP cements to control the setting reaction or limit pH variations.

The hardening reactions probably involve nanocrystal surface interactions between these particles in addition to the conversion of the precursors into apatite. Similar phenomena have been described in biological tissues and are referred to as "crystal fusion".

Table 7. Composition of biomimetic calcium phosphate cements.

Acid-base cements		
Composition	Final product	References
TTCP and DCPD	Hydroxyapatite (Bonesource®)	[Brown 1987]
$CaCO_3$+MCPM and α-TCP	Carbonated apatite (SRS®)	[Constantz 1995]
TTCP + H_3PO_4 and α-TCP	PCA (Cementek®)	[Hatim 1998]
$Ca(OH)_2$ or TTCP and OCP	PCA	[de Maeyer 2000]
Monocomponent cements		
Composition	Final product	References
amorphous CaP	Carbonated apatite (Biobon® and α-BSM®)	[Lee 2000]
α-TCP		[Ginebra 1997, Kon 1998, Dos Santos 1999]
Hydrolyzable phase-based cements		
Composition	Final product	References
β-TCP and MCPM	DCPD converting into PCA	[Mirtchi 1989]

DCPD: dicalcium phosphate dihydrate, OCP: octacalcium phosphate, MCPM: monocalcium phosphate monohydrate, TCP: tricalcium phosphate, TTCP: tetracalcium phosphate.

Recently, original compositions of calcium carbonate-calcium phosphate mixed cements involving the reaction between hydrogenphosphate and carbonate groups have been developed [Combes 2006]. Depending on the initial carbonate/phosphate ratio, these cements can lead to a highly-carbonated nanocrystalline apatite (around 10% w/w of CO_3) analogous to bone mineral associated with a metastable calcium carbonate (vaterite) or not.

6.B. Low-Temperature Sintering of Nanocrystalline Apatites

As a preliminary, we can specify what is considered as "low temperature", since this appreciation appears to be completely dependent on the scientific context. Although ceramists, for example, generally consider temperatures below 1000 °C as being low, the notion of low temperature in this chapter will be reserved to heat treatments enabling the apatite samples to remain nanosized and hydrated, therefore typically in the range 20-300 °C. In these conditions, the term "consolidation" is then probably more suited than "sintering" to describe the physical phenomena undergone by the initial powder samples during the heat treatment.

Calcium phosphate ceramics exhibit excellent biocompatibility and show osteoconductive properties. As such, they are widely used as bone substitute materials and replace autologous grafts or allografts. However, these ceramics are generally obtained by sintering at high temperature and they therefore exhibit very limited surface reactivity and properties. As mentioned above, nanocrystalline apatites offer interesting possibilities: they are the main constituent of bone and nature makes good use of the particularities of their surface and bulk reactivity [Cazalbou 2004]. They are also considered to be at the origin of the biological activity of orthopaedic materials (bioglasses, polymers, ceramics or cements). However their processing while preserving their nanocrystalline nature is a major difficulty in view of the development of bioceramics made of apatite nanocrystals.

In this context, several explorative methods have been attempted such as low temperature hardening from gels, associations with macromolecules or cements [Knaack 1998, Sarda 1999]. However in these cases the mechanical properties observed were too weak to allow a use in load bearing bone sites.

An interesting feature of apatite nanocrystals, as explained in a previous section, is the presence of a surface hydrated layer, and Banu et al. recently investigated the ability of this hydrated layer to favor crystal-crystal interactions [Banu 2005]. They open interesting perspectives in materials science, since ceramic-like materials can be obtained upon drying aqueous gels of nanocrystals [Sarda 1999]. This process is however accompanied by strong shrinkage, which introduces strains in the materials leading to poor mechanical properties. The possibility to use the ion mobility of the hydrated layer to consolidate a nanocrystal assembly has however been established. Recently solid ceramic-like materials have been obtained at very low temperature by low temperature uni-axial pressing [Banu 2005]. The materials obtained at 200°C compared well, regarding mechanical properties, with those obtained by traditional sintering at much higher temperatures (1100-1250 °C), despite a lower densification ratio. At this low temperature slight crystal growth of the apatite domains occurred associated with hydroxylation and a partial decomposition into anhydrous dicalcium phosphate (DCPA).

Beside conventional sintering techniques and hot pressing, spark plasma sintering (SPS) is a technique recently used for the sintering of various kinds of materials at temperatures generally lower than usual, due to the non-conventional heating source (electrical current passing through a conducting matrix). Only few studies of apatite sintering using SPS have been performed, and most of them were done at relatively high temperature (900-1100 °C), not suited to the conservation of the nanosize of apatite crystals analogous to bone mineral. In these reports, SPS sintering was shown to be more efficient when HA powders with a small particle size were used, and the compacts exhibited excellent mechanical properties [Gu 2002]. The formation of transparent apatite ceramics with interesting surface properties has also been reported and generally SPS-sintered HA show greater surface reactivity in SBF (simulated body fluid) tests, often considered as an measurement of the biological activity [Nakahira 2002, Kawagoe 2003].

The process of consolidation of apatite nanocrystals at very low temperature using SPS and the specificity of apatite nanocrystals has not yet been studied in detail. Preliminary results confirm however the possibility to consolidate efficiently such nanocrystals at much lower temperatures than traditional sintering with limited degradation, which opens new fields in the making of ceramic-like biomaterials inspired by the bone mineral using the surface properties of hydrated nanocrystals, and more specifically the high ionic mobility within the surface hydrated layer present on the nanocrystals [Drouet 2006].

For example, consolidation by SPS of biomimetic nanocrystalline apatites matured for one day was investigated, and the spontaneous densification of the powder started around 150 °C and ranged up to 190 °C [Drouet 2006]. Disks obtained by SPS at 180 °C did not exhibit sufficient mechanical resistance, in contrast to those obtained around 200 °C. For the latter, the tensile stress measured by diametral compression test, or "Brazilian" test (adapted to the disk shape) ranged from 18 to 25 MPa for a heating time of 3 minutes from room temperature to 200 °C followed by a plateau at 200 °C for 2 minutes. These values are close to those obtained with stoichiometric hydroxyapatite using the same mechanical test procedure. The corresponding relative density was ca. 70%. Preliminary results also point out the correlated effect of temperature and mechanical pressing for SPS consolidation of nanocrystalline apatites, and a high apparent cohesion of the original agglomerates can be observed by scanning electron microscopy (SEM), which is in agreement with the relatively good mechanical properties.

The XRD pattern of the consolidated disks obtained by SPS treatment for 2 minutes at 200 °C corresponds to a poorly crystalline apatite, although an increase in the degree of crystallinity after SPS treatment is noted. No other crystalline phase is discerned. The average crystallite dimensions were estimated applying Scherrer's formula to lines (002) and (310), leading to a length close to 260 Å and a width of about 80 Å as compared to 220 and 55 Å, respectively, for the starting powder (see section 3.B). These trends point out an increase of the crystallite sizes due to SPS treatment, however the increase of the crystallite length is noticeably lower than that observed after hot pressing treatment, even at low temperature: 295 Å for a treatment at 200 °C for 15 min. The SPS process therefore only causes limited alteration of the initial nanocrystalline powder.

FTIR analysis of SPS consolidated nanocrystalline apatites indicates partial loss of water and the appearance of sharp absorption bands characteristic of apatitic OH^- vibrations. The formation of OH^- ions can be interpreted by the internal hydrolysis of some PO_4^{3-} ions as suggested by Heughebaert [Heughebaert 1982]:

$$PO_4^{3-} + H_2O \rightarrow HPO_4^{2-} + OH^- \qquad \text{(eq. 6)}$$

However, the water loss observed after SPS consolidation is less than that observed after hot pressing processes, even at low temperature (200 °C for 15 min). Also, secondary phases (e.g. pyrophosphates) sometimes observed after hot pressing are not observed by FTIR spectroscopy after SPS treatment [Banu 2005]. This phase purity could represent another advantage of SPS consolidation of nanocrystalline apatites.

6.C. Apatite Coatings

The main subject of concern with the well-known metallic prostheses is the interface between the surrounding bone and the implant surface. During the last two decades various coating methods leading to hydroxyapatite (HA)-coated prostheses combining the good mechanical properties of metals with the excellent biocompatibility and bioactivity of calcium phosphate have been studied. HA-coating is known to improve bone formation when in contact with bone tissue (osteoconduction) and to facilitate the anchorage between the bone tissue and the prosthesis (biointegration). Currently, HA plasma-spraying is the most common and successful process to enhance the bioactivity of metallic implants (orthopaedic and dental implants) and to improve early bone-implant bonding. However, despite its clinical success, the plasma spray process is limited by intrinsic drawbacks: a) the high temperatures involved in this process restrict the process to the deposition of stable phases like stoichiometric hydroxyapatite presenting a low specific surface area (around 1 m^2/g) and thus a low reactivity without any possibility to associate biologically active (macro)molecules (growh factors for example) with the coating, b) the limit of this technique to cover complex surfaces and/or inside porous materials, c) the partial and superficial decomposition of HA particles into several possible phases including CaO, tricalcium phosphate, tetracalcium phosphate and oxyapatite at very high temperatures which could modify the chemical, mechanical and biological behavior of the biomaterial.

Other processes have been studied to improve the quality of the coatings (electrophoretic deposition, ion sputtering deposition and sol-gel methods for example) but in all cases the phase(s) composition of the coating is quite different from biological apatites and the coating reactivity is limited. For example, even though promising cell culture results were obtained (improved cell differentiation compared to pure titanium), one of the main problems of the apatite coating obtained by sol-gel route is the aging time of the precursor sol which can influence the phase composition, homogeneity and textural properties of the deposit [Kim 2004, Izquierdo-Barba 2004].

Recently, other deposition processes involving calcium phosphate supersaturated solution(s) at low temperature have been studied. Several authors explored the concept of biomimetic coatings which consists in soaking metal or polymer implants at physiological pH and temperature in Simulated Body Fluids solution (SBF) mimicking the inorganic ion composition of human blood plasma [Kokubo 1996, Barrère 1999, Leonor 2004]. The main advantages of these processes based on the nucleation and growth of calcium phosphate on metallic substrates from supersaturated solution arise from a) the low processing temperature, b) the formation of a non-stoichiometric apatite quite analogous to bone mineral, c) the possibility to form deposits even on complex implant geometries and to incorporate biologically-active molecules or ions in the coating. However, several drawbacks make this technique difficult to apply on an industrial scale: a) the long immersion time (problem of maturation of apatite and thus a decrease of its reactivity), b) the thinness of the apatite layer deposited and c) the metastability of SBF solutions, and more generally of calcium phosphate supersaturated solutions, which require replenishment and a constant pH to maintain the level of supersaturation for apatite precipitation. Habibovic et al. reported the deposition of an AB type carbonated hydroxyapatite involving two steps using two modified-SBF solutions: firstly, the heterogeneous nucleation of a thin and amorphous CaP layer on the metal surface and secondly the growth of a thick well-crystallized apatite coating [Habibovic 2002]. They have shown that drugs and/or growth factors could be easily incorporated into the biomimetic coating during processing in order to improve the biological performance of the biomaterial (prevention of local infection, controlled drug delivery, enhancement of bone healing).

To avoid the use of supersaturated solutions in industrial processing, another process involving the slow hydrolysis of apatite precursors such as amorphous calcium phosphate first deposited on the metallic surface has been set up [Cazalbou 2002].

On a hydrophilic surface such as the hydrated titanium oxide-hydroxide layer at the surface of a titanium implant in aqueous media, the interaction between apatite nanocrystals and the titanium surface can be achieved through the hydrated surfaces interacting with a bonding process as illustrated in figure 15. The labile ions from the superficial hydrated layer of freshly formed apatite allow a topological fitting to the metallic surface. The progressive transformation of the superficial layer into regular apatite structure improves the adhesion of the nanocrystals to the surface. Upon drying strong bonding may be formed by direct ionic interactions. Even though no systematic studies have been reported, several parameters involved in the quality of the bonding can be distinguished: the drying rate and/or the rate of evolution of the hydrated layer toward regular apatite structure.

Another application that takes advantage of the PCA reactivity is the activation of three-dimensional porous scaffolds often used for bone regeneration (bone filling and substitute, tissue engineering) by coating their accessible surfaces with biomimetic nanocrystalline apatites analogous to bone mineral. Indeed, although generally based on calcium phosphate phases (e.g. biphasic calcium phosphate ceramics (BCP): HA/β-TCP), the processing of such scaffolds usually leads to a strong decrease of their bioactivity due to high-temperature treatments and the coating of the ceramic open pores with PCA make it possible to take advantage of the physical-chemical properties of such nanocrystalline apatite phase to enhance the ceramic bioactivity. Figure 16 shows a poorly crystalline apatite coating (around 3 µm-thick) obtained by impregnation by PCA gel on the accessible surface of a porous BCP ceramic. Further activation can also be achieved by surface ion exchange between the biomimetic apatite coating and aqueous solutions enriched with mineral ions such as magnesium and carbonate (see in section 5.B).

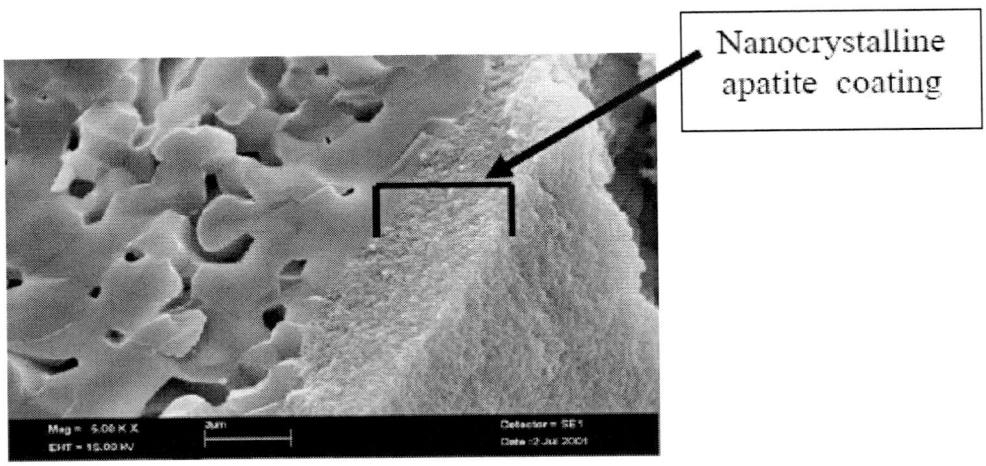

Figure 16. SEM micrograph of nanocrystalline apatite deposit on porous HA/TCP scaffold accessible surface.

6.D. Composites

The model of interactions between platelets of apatite nanocrystals and collagen (protein) occurring in bone tissue, although not yet completely understood, can be exploited to prepare hybrid mineral-organic materials. The association of calcium phosphates with organic molecules (mineral-organic composites) can impart bioactive and mechanical properties closer to those of bone compared to pure calcium phosphate ceramics. The possibility for molecules to interact with the ions from the apatite surface layer through ionic functions of proteins (mostly anionic groups) has already been mentioned in this chapter (see in section 5c). Polymer- or protein-apatite composites can combine a better structural integrity and flexibility along with good bioactivity. Recently, such composites have gained increasing interest in the field of tissue engineering which appears as a promising concept for bone reconstruction. This concept requires a biodegradable host matrix (composite for example) acting temporarily as a mechanical support and directing osteoblast growth and tissue neoformation once implanted. High porosity is necessary to promote implant binding and integration and especially neoformed bone invasion.

Generally, composite materials (ceramic-polymer) are obtained with classical mixing techniques of constituents but other techniques inspired by in vivo hard tissue calcification processes have emerged. For example, the use of organized organic matrices with numerous sites favourable for nucleation of calcium phosphates after phosphatation (of cellulose for example) or sodium silicate treatments [Miyaji 1999] .

Many studies can be found in the literature pursuing the aim to produce biomimetic artifcial bone-like tissue involving HA and collagen as fiber, gel or gelatin (denatured collagen) [Itoh 2000, Tampieri 2003, Kikuchi 2004, Kim 2005]. Testing two methods of preparation of apatite/collagen composite materials (dispersion of HA in collagen gel or direct nucleation of HA into collagen fibers), Tampieri et al. have shown that the bio-inspired method based on the direct nucleation of apatite leads to composites analogous to calcified tissue and exhibiting strong interactions between HA and collagen [Tampieri 2003].

In the dog, in vivo evaluation in weight-bearing sites testing apatite/collagen composites prepared by co-precipitation method has shown interesting results especially when associated with recombinant human bone morphogenetic protein 2 (rhBMP-2) [Itoh 2002]. The controlled release of rhBMP-2 from the implant facilitates early formation of calli and hence new bone enabling early weight bearing.

Nishikawa et al. have examined the biodegradation of HA/collagen composites implanted in dogs by a tissue labeling method; they showed that this novel composite is useful for bone augmentation and that the calcium in the newly formed bone might have been released from the implant [Nishikawa 2005].

The association of apatite with chitosan (a natural polysaccharide obtained by deacetylation of chitin) is expected to give interesting composites for tissue engineering applications due to the biocompatibility, biodegradability and bioactivity properties of chitosan. Several authors have reported the preparation of porous chitosan-apatite composites using co-precipitation methods [Kong 2005, Rusu 2005]. The association of chitosan with apatite improved the biocompatibility of the material and greater cell proliferation on a composite scaffold has been reported [Kong 2005].

In our research group, protein-CaP associations have been prepared according to biomimetic processes at low temperature and in aqueous suspension. The proteins studied (casein, albumin) present an affinity for calcium phosphate surfaces. After evaporation of water (according to figure 15 drying scheme, "low temperature sintering" process) in the protein-CaP suspension, nanocrystalline apatite associated with protein gives a bulk composite (protein-nanocrystalline apatite) with low porosity but with poor mechanical properties compared to those of bone [Sarda 1999].

Macroporous composite scaffolds can be processed using different methods: solvent casting/particulate leaching, emulsion freeze drying or thermally induced phase separation.

Biodegradable composite including resorbable polymer such as polylactic acid (PLA) and/or polyglycolic acid (PGA) and a resorbable apatite can be prepared at ambient temperature [Linhart 2001, Chen 2001]. This kind of association has the advantage of a controlled pH in the surrounding medium during the composite degradation since apatite (whose degradation involves the release of alkaline components) can moderate the pH drop resulting from polyester (PLA or PGA) biodegradation (hydrolysis of ester bond). Other good reasons for using such synthetic or natural polymers are: a) the ease of processing , b) the possibility to control the polymer degradation rate depending on its composition (polymer or copolymer), molecular weight and crystallinity. Moreover, the degradation in vivo of PLA releases lactic acid, which is a natural metabolite and thus remains under the control of the organism. The biomaterial structure (dense or porous composite) also has an effect on the rate of biodegradation which is lower in the case of porous composites where the acid released from PLA degradation can be leached with biological fluids whereas this acid concentrated in the bulk of dense composite should autocatalyze composite biodegradation.

Recently, Mathieu et al. reported the use of a supercritical CO_2 foaming process to prepare porous PLA-HA and/or PLA-TCP composites exhibiting mechanical behavior analogous to bone (anisotropy in compressive and viscoelastic properties) [Mathieu 2006].

Finally, the formation of a polymer-apatite composite can also correspond to a first step in the preparation of nanocrystalline apatite porous ceramic. For example, Tadic et al. reported the preparation of nanocrystalline apatite-based porous bioceramics using both sodium chloride salt and polyvinyl alcohol fibers as water-soluble porogenic agents and cold isostatic pressing without the need to sinter [Tadic 2004].

7. Biological Properties of Nanocrystalline Apatites

Three main characteristics have to be considered: biodegradation, cell-materials interactions and materials-tissue interactions.

7.A. Biodegradation

The main advantage of apatite nanocrystals-based biomaterials is their similarity with bone mineral. However these nanocrystals may have very different characteristics which account for the large differences in resorption rates. Several factors involved in biodegradation have been identified.

The composition of CaP nanocrystals appears as an important factor. Resorption is generally believed to be related to the solubility of CaP materials [Legeros 1993]. Considering nanocrystalline apatites, the solubility depends on the presence of vacancies and on the molar Ca/(P+C) ratio (where P represents phosphate ions and C the carbonate ions). Low Ca/(P+C) ratios indicate a high amount of cationic vacancies and a less cohesive solid with a higher solubility. Non-stoichiometric apatites, however, have the ability to mature like bone mineral and may evolve toward more stable, less soluble compounds upon aging. Thus a change in solubility may occur with time and the rate of bioresorption may decrease. These evolutions and alterations of the nanocrystal composition with implantation time have not been studied extensively, but will be crucial for understanding the degradation behavior of the nanocrystals.

The amount of labile ("non-apatitic") species may also considerably modify the dissolution characteristics and thus the resorption rate. It is likely that the labile species determine the "abnormal" dissolution characteristics of bone crystals [Baig 1999]. These factors depend on the composition of the materials and on their conditions of formation. Some materials, involving inhibitors of crystal maturation such as carbonate, magnesium or pyrophosphate, consist of small apatitic crystals rich in labile non-apatitic environments that will be rapidly resorbed. Others contain bigger, more mature crystals with a slower resorption rate. The presence of alkaline residues or the existence of alkaline reactions on contact with body fluids may strongly disturb biodegradation by cells as slight alkalinity was shown to greatly reduce osteoclast resorption [Kim 2001].

In addition to specific characteristics of the nanocrystals, their bonding scheme and the characteristics of the materials (porosity, size, presence of other constituents) also play a considerable role in the biodegradation process as for other types of biomaterials.

Although composed of apatite nanocrystals analogous to bone mineral, the degradation of biomimetic apatite-based materials may also produce small particles that can induce a local inflammatory response.

7.B. Biological Activity

The origin of biological activity has not really been investigated in all cases. Most studies concern implantations in animals, but the behavior of osteoblasts on CaP biomimetic materials is not well known. Several studies performed for example on cement particles have shown an adverse effect on osteoblast viability, proliferation and synthesis of extracellular matrix [Pioletti 2000]. These effects however, are not due to the cement itself but rather to the presence of loose particles and in fact may be observed with any kind of material releasing small particles. Generally, all implantation tests of biomimetic CaP materials have given excellent results. There is generally no formation of fibrous tissue at the contact of the implant, and the material is directly attached to bone. In most cases, the irregular resorption zone is invaded by new forming bone and osteons can often be distinguished as in regular bone remodeling. A few islands of materials are imbedded in new forming bone at the end of the process [Frankenburg 1998, Knaack 1998, Yuan 2000].

Some biomimetic cements and coatings have been reported to be osteoinductive. However the manifestations of this behavior seem rather erratic. Although assigned to microporosity and the preferential uptake of circulating growth factors, the origin of this behavior needs to be demonstrated and clarified [Habibovic 2006].

The reasons for biological activity and its correlation with the characteristics of the biomimetic material are difficult to extract from existing data. Nevertheless knowing which biomaterial-tissue interactions occur allows a probable scenario to be sketched. The bioactivity of orthopaedic materials has been shown to result from the precipitation of a neoformed layer of carbonated apatite crystals analogous to bone mineral crystals in vivo, which enables osteoblast cell adhesion and proliferation. The formation of this neo-formed apatite layer can be induced either by nucleation of crystals from supersaturated biological fluids or by induction of nucleation resulting from increases in the local concentrations of Calcium, phosphate, carbonate or hydroxide ions. The first process might not necessarily occur in the presence of non-stoichiometric apatite nanocrystals which show a higher solubility than stoichiometric apatite and would be in quasi-equilibrium with body fluids. However the nanocrystals with their very high specific surface area offer numerous crystal growth sites which can promote the formation of neoformed crystals. The second process may be induced by the presence of other components. In several cases unstable phases existing in the materials may play a crucial role by locally increasing the mineral ion concentrations favoring additional precipitation of very reactive apatite from body fluids. These neoformed crystals and/or the exposed biomimetic nanocrystals of the material itself exhibit a very high specific surface area and a high surface reactivity due to the presence of labile, non-apatitic environments of mineral ions like in freshly formed bone crystals. They can easily bind proteins including growth factors which can favor cell attachment, proliferation and activity. The behavior of biomimetic materials may thus be rather complex. It can also vary with the time of implantation, the porosity and the size of the specimen. Proteins and cells will not diffuse through long distances. As maturation of the nanocrystalline apatitic phases progresses, biological activity, like resorption ability, can vary with time and with the size of the bone defect. These phenomena however, may also be true for other types of ceramic implants and are not specific to the behavior of biomimetic materials. Several additives may be used to delay maturation and preserve the activity of inner surfaces.

7.C. Materials-Tissue Interactions and Biointegration

In addition to tissue reconstruction around and within implants, biomaterials can bind directly to living tissue without the interposition of a fibrous layer. The bone-bonding ability is determined by mechanical interlocking and chemical interactions. The first effect is related to the general geometry of the samples and has little to do with its composition. Chemical interactions are believed to mainly involve the neo-formed apatite layer, however in the case of biomimetic nanocrystals the surface characteristics are already similar to that of bone tissue. Although bone bonding processes are not yet clearly known, it seems that the junction zone between the implant and bone tissue shows similar characteristics to the cement line separating old osteons from new osteons after bone remodelling: a disordered area with specific proteins such as ostepontin and bone sialoproteins [Kawagushi 1993, Davies 2003]. Thus, the bonding of living bone to biomimetic nanocrystalline apatite materials could be very similar to that occuring in bone remodelling processes. This binding process could explain the strength of the bone-apatite biomaterials interface. The interactions at the molecular level are not yet precisely known but it seems probable that the very reactive surface of bone nanocrystals could be involved in interactions both with the mineral fraction and the organic fraction.

Conclusion

Biomimetic nanocrystalline apatites exhibit enhanced and tunable reactivity as well as original surface properties related to their composition and mode of formation. Synthetic nanocrystalline apatites analogous to bone mineral can be easily prepared in aqueous media and one of the most interesting characteristics of their nanocrystals is the existence of a hydrated surface layer containing labile ionic species. The latter can be easily and rapidly exchanged with ions and/or macromolecules from the surrounding fluids. Even though the surface composition and structure of apatite nanocrystals have not yet been completely unveiled, the fine characterization of these very reactive nanocrystals can be investigated in different ways including chemical analysis, X-ray diffraction and spectroscopic techniques such as FTIR and solid state NMR. All these investigations are also essential to throw light on bone mineral reactivity and properties in general.

The ion mobility in the hydrated layer allows direct crystal-crystal or crystal-substrate bonding offering material scientists and medical engineers extensive possibilities for the design of bone substitute materials with improved bioactivity using unconventional processing. Indeed apatitic biomaterials can be processed in different forms (dense or porous ceramics, cements, composites and coatings) at low temperatures, which preserves their surface reactivity and biological properties. They can also be associated with bioactive (macro)molecules and/or ions to improve the biological performance of bone substitute materials.

REFERENCES

Arsenault A.L., Grynpas M.D. (1988). Crystals in calcified epiphyseal cartilage and cortical bone of the rat. *Calcified Tissue International*, 43, 219-225.

Aue W.P., Roufosse A.H., Glimcher M.J., Griffin R.G. (1984). Solid state phophorus-31 Nuclear Magnetic Resonance of calcium phosphate: Potential models of bone mineral. *Biochemistry,* 23, 6110-6114.

Baig A.A., Fox J.L., Hsu J., Wang Z., Otsuka M., Higuchi W.I., Legeros R.Z. (1996). The effect of carbonate content and crystallinity on the metastable equilibrium solubility behavior of carbonated apatites. *J Colloid Interface Sci*., 179, 608–617.

Baig AA, Fox JL, Wang Z, Higuchi WI, Miller SC, Barry AM, Otsuka M. (1999). Metastable equilibrium solubility behavior of bone mineral. *Calcif. Tissue Int.,* 64, 329-339.

Baig A.A., Fox J.L., Young R.A., Wang Z., Hsu J., Higuchi W.I., Chhettry A., Zhuang H., Otsuka M. (1999). Relationships among carbonated apatite solubility, crystallite size, and microstrain parameters. *Calcified Tissue International*, 64, 437–449.

Banu M. Ph.D Thesis, INPT, France, 2005.

Baroud G., Swanson T., Steffen T. (2006). Setting properties of four acrylic and two calcium phosphate cements used in vertebroplasty. *J. Long Term Eff. Med. Implants*, 16, 51-59.

Barrère F., Layrolle P., van Blitterswijk C.A., de Groot K. (1999). Biomimetic calcium phosphate coatings on Ti6AI4V: a crystal growth study of octacalcium phosphate and inhibition by Mg^{2+} and HCO_3^-. *Bone*, 25, 107S-111S.

Belton P.S., Harris R.K., Wilkes P.J. (1988). Solid state phophorus-31 NMR studies of synthetic inorganic phosphates. *J. Phys. Chem. Solids*, 49, 21-27.

Benaziz L., Barroug A., Legrouri A., Rey C., Lebugle A. (2001). Adsorption of O-phospho-L-serine and L-serine onto poorly crystalline apatite. *J. Colloid and Interface Science*, 238, 48-53.

Bennani-Ziatni M., Lebugle A., Taitai A., Ferhat M. (1997). Adsorption of glycine from a water-ethanol medium onto precipitated calcium phosphates. *Annales de Chimie,* 22, 537-547.

Bennani-Ziatni M., Lebugle A., Taitai A. (2003). Influence of dielectric constant on the adsorption of glycine by bioapatites. *Phosphorus, Sulfur and Silicon and the Related Elements*, 178, 221-233.

Beshah K., Rey C., Glimcher M.J., Schimizu M., Griffin R.G. (1990). Solid state carbon-13 and proton NMR studies of carbonate-containing calcium phosphates and enamel. *J. Solid State Chem*., 84, 71-81.

Bihi N., Bennani-Ziatni M., Taitai A., Lebugle A. (2002). Adsorption of amino-acids onto bone-like carbonated calcium phosphates, *Annales de Chimie*, 27, 61-70.

Bonar L.C., Roufosse A.H., Sabine W.K., Grynpas M.D., Glimcher M.J. (1983). X-ray diffraction studies of the crystallinity of bone mineral in newly synthesized and density fractionated bone. *Calcified Tissue International,* 35, 202-209.

Bonfield W. (1988). Composites for bone replacement. *J Biomed Eng*, 10, 522-526.

Brown W.E., Chow L.C. (1987). A new calcium phosphate water-setting cement. In P Brown (Ed.), Cement Research Progress 1986, pp. 351-379. Westerville OH: *American Ceramic Society.*

Brown W.E., Eidelman N., Tomazic B. (1987). Octacalcium phosphate as a precursor in biomineral formation. *Adv. Dental Res*., 1, 306-313.

Burnell J.M., Teubner E.J., Miller A.G. (1980). Normal maturational changes in bone matrix, mineral and crystal size in the rat. *Calcified Tissue International*, 31, 13-19.

Campbell A.C., LoRe M., Nancollas G.H. (1991). The influence of carbonate and magnesium ions on the growth of hydroxyapatite and carbonated apatite and human powdered enamel. *Colloids and Surfaces* A, 54, 25-31.

Cazalbou S., Hina A., Rey C. (1999). Interactions between trace elements and bone mineral matrix. In M. Abdulla, M. Bost, S. Gamon, P. Arnaud, G. Chazot (Eds.*), New Aspects of Trace Element Research*, pp. 58-62. London, Smith-Gordon.

Cazalbou S., Ph.D. Thesis, INPT, France, 2000.

Cazalbou S., Combes C., Rey C. (2002). Procédé permettant de recouvrir à basse température des surfaces par des phosphates apatitiques nanocristallins à partir d'une suspension aqueuse de phosphate amorphe. Patent n° FR0209514.

Cazalbou S., Combes C., Eichert D., Rey C., Glimcher M.J. (2004). Poorly crystalline apatites: evolution and maturation in vitro and in vivo. *J Bone Miner Metab*, 22, 310-317.

Cazalbou S., Combes C., Eichert D., Rey C. (2004). Adaptative physico-chemistry of bio-related calcium phosphates. *Journal of Materials Chemistry*, 14, 2148-2153.

Chen G., Ushida T., Tateishi T. (2001). Poly(DL-lactic-co-glycolic acid) sponge hybridized with collagen microsponges and deposited apatite particulates. *J Biomed Mater Res*, 57, 8-14.

Cho G., Yesionowski J.P. (1996). ^{1}H and ^{19}F Multiple-Quantum NMR dynamics in quasi-one-dimensional spin clusters in apatites. *J. Phys. Chem*., 100, 15716-15725.

Cho G., Wu Y., Ackerman J.L. (2003). Detection of hydroxyl ions in bone mineral by solid-state NMR spectroscopy. *Science,* 300, 1123-1127.

Combes C., Rey C., Mounic S. (2001). Identification and evaluation of HPO_4 ions in biomimetic poorly crystalline apatite and bone mineral. *Key Engineering Materials*, vol 192-195, 143-146.

Combes C., Rey C. (2002). Adsorption of proteins and calcium phosphate materials bioactivity. *Biomaterials*, 23, 2817-2823.

Combes C., Bareille R., Rey C. (2006). Calcium carbonate-calcium phosphate mixed cement compositions for bone reconstruction. *J Biomed Mater Res A* (in press).

Constantz B.R., Ison I.C., Fulmer M.T., Poser R.D., Smith S.T., VanWagoner M., Ross J., Goldstein S.A., Jupiter J.B., Rosenthal D.I. (1995). Skeletal repair by in situ formation of the mineral phase of bone. *Science,* 267, 1796-1798.

Dabbarh F., Lebugle A., Taitai A., Bennani M. (2000). Influence du séchage sur la composition de phosphates de calcium carbonatés, analogues osseux. *Ann Chim. Sci. Mat.*, 25, 339-348.

Daculsi G., Menanteau J., Kerebel L.M., Mitre D. (1984). Length and shape of enamel crystals. *Calcif Tissue Int,* 36, 550-555.

Daculsi G., Laboux O., Malard O., Weiss P. (2003). Current state of the art of biphasic calcium phosphate bioceramics. *J. Mater. Sci. Mater. Med.*, 14, 195-200.

Danilchenko S.N., Kukharenko O.G., Moseke C., Protsenko I.Y., Sukhodub L.F., Sulkio-Cleff B. (2002). Determination of the bone mineral crystallite size and lattice strain from diffraction line broadening. *Cryst. Res. Technol.*, 37, 1234-1240.

Davies J.E. (2003). Understanding peri-implant endosseous healing. *J. Dent. Educ.*, 67, 932-949.

de Groot K., Geesink R., Klein C.P., Serekian P. (1987). Plasma sprayed coatings of hydroxyapatite. *J. Biomed. Mater. Res.,* 21, 1375-1381.

de Jong W.F. (1926). The mineral substance in bones. *Rec. Trav. Chim., Pays-Bas*, 45, 445-448.

de Maeyer E.A., Verbeeck R.M., Vercruysse C.W. (2000). Conversion of octacalcium phosphate in calcium phosphate cements. *J Biomed Mater Res*, 52, 95-106.

dos Santos L.A., de Oliveira L.C., Rigo E.C., Carrodeguas R.G., Boschi A.O., De Arruda A.C. (1999). Influence of polymeric additives on the mechanical properties of α-tricalcium phosphate cement. *Bone*, 25, 99S-102S.

Drouet C., Largeot C., Raimbeaux G., Estournès C., Dechambre G., Combes C., Rey C. (2006). Bioceramics: spark plasma sintering (SPS) of calcium phosphates. *Adv Sci Tech*, 49, 45-50.

Eichert D. Ph.D. Thesis, INPT, France, 2001.

Eichert D., Sfihi H., Combes C., Rey C. (2004). Specific characteristics of wet nanocrystalline apatites. Consequences on biomaterials and bone tissue. *Key Engineering Materials,* vol. 254-256, 927-930.

Eichert D., Salomé M., Banu M., Susini J., Rey C. (2005). Preliminary characterization of calcium chemical environment in apatite and non-apatitic calcium phosphates of biological interest by X-ray absorption spectroscopy. *Spectrochimica Acta* B, 60, 850-858.

Elliott J.C., Mackie P.E., Young R.A. (1973). Monoclinic hydroxyapatite. *Science,* 180, 1055-1057.

Elliott J.C., Holcomb D.W., Young R.W. (1985). Infrared determination of the degree of substitution of hydroxyl by carbonate ions in human dental enamel. *Calcif. Tissue Int.*, 37, 372-375.

Elliott J.C. (1994). Structure and Chemistry of the Apatites and Other Calcium Orthophosphates. Amsterdam, Elsevier Science BV.

Fernandez Gavarron F. (1978). The dynamic equilibrium of calcium. In E. Pina, A. Pena, V. Chagoya de Sanchez (Eds.), *Temas Bioquim.* Actual., pp. 41-67. Mexico Cit, Univ. Nac. Auton. Mexico.

Fisher L.W., Eanes E.D., Denholm L.J., Heywood B.R., Termine J.D. (1987). Two bovine models of osteogenesis imperfecta exhibit decreased apatite crystal size. *Calcified Tissue International*, 40, 282-285.

Frankenburg E.P., Goldstein S.A., Bauer T.W., Harris S.A., Poser R.D. (1998). Biomechanical and histological evaluation of a calcium phosphate cement. *J Bone Joint Surg Am*, 80, 1112-1124.

Gee A., Dietz V.R. (1953). Determination of phosphate by differential spectrophotometry. *Ann. Chem.,* 25, 1320-1324.

Gee A., Dietz V.R. (1955). Pyrophosphate formation upon ignition of precipitated basic calcium phosphate. *J. Am. Chem. Soc.*, 77, 2961-2965.

Ginebra M.P., Fernandez E., De Maeyer E.A., Verbeeck R.M., Boltong M.G., Ginebra J., Driessens F.C., Planell J.A. (1997). Setting reaction and hardening of an apatitic calcium phosphate cement. *J Dent Res*, 76, 905-912.

Glimcher M.J. (1990). The possible role of collagen fibrils and collagen-phosphoprotein complexes in vitro and in vivo. *Biomaterials,* 11, 7-10.

Grynpas M.D., Bonar L.C., Glimcher M.J. (1984). X-ray diffraction radial distribution function studies of bone mineral and synthetic calcium phosphate. *J. Mater. Sci.*, 19, 723-736.

Gu Y.W., loh N.H., Khor K.A., Tor S.B., Cheang P. (2002). Spark plasma sintering of hydroxyapatite powders. *Biomaterials*, 23, 37-43.

Habibovic P., Barrère F., van Blitterswicjk C.A., de Groot K., Layrolle P. (2002). Biomimetic hydroxyapatite coating on metal implants. *J Am Ceram* Soc, 85, 517-522.

Habibovic P., Sees T.M., van den Doel M.A., van Blitterswijk C.A., de Groot K. (2006). Osteoinduction by biomaterials. Physicochemical and structural influences. *J. Biomed. Mater. Res.* A, 77, 747-762.

Haris Parvez I., Bigi A., Boanini E., Gazzano M., Rubini K., Torricelli P. (2004). Nanocrystalline hydroxyapatite-polyaspartate composites. *Bio-medical Materials and Engineering*, 14, 573-579.

Hatim Z., Frèche M., Kheribech A., Lacout J.L. (1998). The setting mechanism of a phosphocalcium biological cement. *Ann. Chim. Sci. Mat.,* 23, 65-68.

Heughebaert J.C., Montel G. (1982). Conversion of amorphous tricalcium phosphate into apatitic tricalcium phosphate. *Calcif Tiss Int,* 34, S103-S108.

Hina A. Ph.D. Thesis, INPT, France, 1996.

Huffman E.W.D. (1977). Performance of a new automatic carbon dioxide coulometer. *Microchemical J,* 22, 567-573.

Ishikawa K., Ducheyne P., Radin S. (1993). Determination of the calcium/phosphorus ratio in calcium-deficient hydroxyapatite using X-ray diffraction analysis. *J. Mat. Sci: Mater*. In Med., 4, 165-168.

Isobe T. , Nakamura S., Nemoto R., Senna M., Sfihi. H. (2002). Solid state double nuclear magnetic resonance of calcium phosphate nanoparticules synthesized by wet-mechanochemical reaction. *J. Phys. Chem* B, 106, 5169-5176.

Itoh S., Kikuchi M., Takakuda K., Koyama Y., Matsumoto H.N., Ichinose S., Tanaka J., Kawauchi T., Shinomiya K. (2000). The biocompatibility and osteoconductive activity of a novel hydroxyapatite /collagen composite biomaterial, and its function as a carrier of rhBMP-2. *J. Biomed Mater Res*, 54, 445-453.

Itoh S., Kikuchi M., Takakuda K., Nagaoka K., Koyama Y., Tanaka J., Shinomiya K. (2002). Implantation study of a novel hydroxyapatite /collagen (Hap/col) composite into weight-bearing sites of dogs. *J. Biomed Mater Res*, 63, 507-515.

Iyengar G.V., Tandon L. Minor and trace elements in human bones and teeth. *International Atomic Energy Agency, NAHRES-39 report*, Vienna,1999.

Izquierdo-Barba I., Hijon N., Cabanas M.V., Vallet-Regi M. (2004). Apatite layers by a sol-gel route. *Key Engineering Materials,* vol. 254-256, 363-366.

Jäger C., Welzel T., Meyer-Zaika W., Epple M. (2006). A solid state NMR investigation of the structure of nanocrystalline hydroxyapatite. *Magn. Reson. Chem.*, 44, 573-580.

Jarcho M., Salsbury R.L., Thomas M.B., Doremus R.H. (1979). Synthesis and fabrication of β-tricalcium phosphate ceramics for potential prosthetic applications. *J Mater Sci,* 14, 142-150.

Jarlbring M., Sandström D.E., Antzutkin O.N., Forsling W. (2006). Characterization of Active Phosphorus Surface Sites at Synthetic Carbonate-Free Fluorapatite Using Single-Pulse ^{1}H, ^{31}P, and ^{31}P CP MAS NMR. Langmuir, 22, 4787 - 4792.

Johnson A.R., Armstrong W.D., Singer L. (1970). Exchangeability of calcium and strontium of bone in vitro. *Calcified Tissue Research,* 6, 103-112.

Kawagoe D., Ioku K., Fujimori H., Goto S. (2003). In vitro estimation with simulated body fluid of OH-designed transparent apatite ceramics prepared by spark plasma sintering. *Trans. Mater. Res. Soc.* Japan., 28, 841-844.

Kawagushi H., McKee M.D., Okamoto H., Nanci A. (1993). Tissue response to hydroxyapatite implanted in an alveolar bone defect: Immunocytochemical and lecitin-gold characterization of the hydroxyapatite-bone interface. *Cells and Materials,* 3, 337-350.

Kikuchi M., Ikoma T., Itoh S., Matsumoto H.N., Koyama Y., Takakuda K., Shinomiya K., Tanaka J. (2004). Biomimetic synthesis of bone-like nanocomposites using the self-organization mechanism of hydroxyapatite and collagen. *Composites Science and Technology,* 64, 819-825.

Kim H.M., Kim Y.S., Woo K.M., Park S.J., Rey C., Kim Y., Kim J.K., Ko J.S. (2001). Dissolution of poorly crystalline apatite crystals by osteoclasts determined on artificial thin-film apatite. *J. Biomed. Mater. Res.,* 56, 250-256.

Kim H.W., Kim H.E., Knowles J.C. (2004). Sol-gel films on titanium implant for hard tissue regeneration. *Key Engineering Materials,* vol. 254-256, 423-426.

Kim H.W., Knowles J.C., Kim H.E. (2005). Porous scaffolds of gelatin-hydroxyapatite nanocomposites obtained by biomimetic approach: characterization and antibiotic drug release. *J Biomed Mater Res,* 74B, 686-698.

Knaack D., Goad M.E.P., Ailova M., Rey C., Tofighi A., Chakravarthy P., Lee D. (1998). Resorbable calcium phosphate bone substitute. *J. Biomed. Mat. Res.*, 43, 399-409.

Kokubo T. (1996). Formation of biologically active bone-like apatite on metals and polymers by a biomimetic process. *Thermochimica Acta*, 280/281, 479-490.

Kon M., Miyamoto Y., Asaoka K., Ishikawa K., Lee H.H. (1998). Development of calcium phosphate cement for rapid crystallization to apatite. *Dent Mater J*, 17, 223-232.

Kong L., Gao Y., Cao W., Gong Y., Zhao N., Zhang X. (2005). Preparation and characterization of nano-hydroxyapatite/chitosan composite scaffolds. *J Biomed Mater Res*, 75, 275-282.

Labarthe J.C., Bonel G., Montel G. (1973). Structure and properties of B-type phosphocalcium carbonate apatites. *Annales de Chimie* (Fr), 8, 289-301.

Lebugle A., Zahidi E., Bonel G. (1986). Effect of structure and composition on the thermal decomposition of calcium phosphates (Ca/P = 1.33). *Reactivity of Solids*, 2, 151-161.

Lee D., Rey C., Aiolova M., Tofighi A. (2000). Methods and products related to the physical conversion of reactive amorphous calcium phosphate, US patent N° 6117456.

Legeros R.Z., Trautz O.R., Legeros J.P., Klein E. (1968). Carbonate substitution in the apatitic structure. Bull. *Soc. Chim. Fr. n° special*, 1712-1718.

Legeros RZ. (1993). Biodegradation and bioresorption of calcium phosphate ceramics. *Clin. Mater.*, 14, 65-88.

Legeros R.Z. (1994). Biological and synthetic apatites. In P.W. Brown, B.Constantz (Eds.), *Hydroxyapatite and Related Materials*, pp. 3-28. Boca Raton, CRC Press.

Legeros R.Z. (2002). Properties of osteoinductive biomaterials: calcium phosphates. *Clin. Orthop. Relat. Res.*, 395, 81-98.

Legros R., Balmain N., Bonel G. (1987). Age-related changes in mineral of rat and bovine cortical bone. *Calcif. Tissue Int.*, 41, 137-144.

Leonor I.B., Azevedo H.S., Pashkuleva I., Oliveira A.L., Alves C.M., Reis R.L. (2004). Learning from nature how to design biomimetic calcium-phosphate coatings. In R.L. Reis and S. Weiner (Eds.), Learning from nature how to design new implantable biomaterials: from biomineralization fundamentals to biomimetic materials and processing routes, Nato Science Series; series II: *Mathematics, Physics and Chemistry* - vol. 171, pp. 123-150. Dordrecht, Kluwer Academic Publishers.

Leung Y., Walters M.A., LeGeros R.Z. (1990). Second derivative infrared spectra of hydroxyapatite. *Spectrochimica Acta A*., 46, 1453-1459.

Linhart W., Peters F., Lehmann W., Schwarz K., Schilling A.F., Amling M., Rueger J.M., Epple M. (2001). Biologically and chemically optimized composites of carbonated apatite and polyglycolide as bone substitution materials. *J Biomed Mater Res*, 54, 162-171.

Liu D.M., Troczynski T., Tseng W.J. (2002). Aging effect on the phase evolution of water-based sol-gel hydroxyapatite. *Biomaterials,* 23, 1227-1236.

Luong L.N., Hong S.I., Patel R.J., Outslay M.E., Kohn D.H. (2006). Spatial control of protein within biomimetically nucleated mineral. *Biomaterials*, 27, 1175-1186.

MacConnel D. (1973). *Apatite, its Crystal Chemistry, Mineralogy, Utilization, and Geologic and Biologic Occurrences.* New York, Spinger-Verlag.

Marie P.J., Ammann P., Boivin G., Rey C. (2001). Mechanisms of action and therapeutic potential of strontium in bone. *Calcified Tissue International*, 69, 121-129.

Mathew M., Brown W.E., Schoeder L.W., Dickens B. (1988). Crystal structure of octacalcium bis(hydrogenphosphate) tetrakis(phosphate) pentahydrate, Ca_8 $(HPO_4)_2$ $(PO_4)_4$, $5H_2O$. *J. Cryst Spectrosc. Res*., 18, 235-250.

Mathieu L.M., Mueller T.L., Bourban P.E., Pioletti D.P., Muller R., Manson J.A.E. (2006). Architecture and properties of anisotropic polymer composite scaffolds for bone tissue engineering. *Biomaterials,* 27, 905-916.

Midy V., Rey C., Brès E., Dard M. (1998). Basic fibroblast growth factor adsorption and release properties of calcium phosphate. *J Biomed Mater Res*, 41, 405-411.

Miquel J.L., Facchini L., Legrand A P., Rey C., Lemaitre J. (1990). Solid state NMR to study calcium phosphate ceramics. *Colloids and Surfaces*, 45, 427-433.

Mirtchi A.A., Lemaitre J., Terao N. (1989). Calcium phosphate cements: study of the beta-tricalcium phosphate-monocalcium phosphate system. *Biomaterials,* 10, 475-480.

Miyaji F., Kim H.M., Handa S., Kokubo T., Nakamura T. (1999). Bonelike apatite coating on organic polymers: novel nucleation process using sodium silicate solution. *Biomaterials*, 20, 913-919.

Nakahira A., Tamai M., Aritani H., Nakamura S., Yamashita K. (2002). Biocompatibility of dense hydroxyapatite prepared using a SPS Process. *J. Biomed. Mater. Res.*, 62, 550-557.

Neuman W.F., Toribara T.Y., Mulryan B.J. (1956). The surface chemistry of bone. IX. Carbonate: phosphate exchange. *J. Am. Chem. Soc.,* 78, 4263-4266.

Neuman W.F., Terepka A.R., Canas F., Triffitt J.T. (1968). Cycling concept of exchange in bone. *Calcified Tissue Research,* 2, 262-270.

Neuman W.F., Neuman M.W. (1985). *Blood: bone calcium homeostasis.* Shika Kiso Igakkai Zasshi, 24, 272-281.

Nishikawa T., Masuno K., Tominaga K., Koyama Y., Yamada T., Takakuda K., Kikuchi M., Tanaka J., Tanaka A. (2005). Bone repair analysis in a novel biodegradable hydroxyapatite/collagen composite implanted in bone. *Implant Dent,* 14, 252-260.

Ouizat S., Barroug A., Legrouri A., Rey C. (1999). Adsorption of bovine serum albumin on poorly crystalline apatite: influence of maturation. *Mat. Res. Bull.,* 34, 2279-2289.

Pak C.Y.C., Bartter F.C. (1967). Ionic interaction with bone mineral. I. Evidence for an isoionic calcium exchange with hydroxyapatite. *Biochimica et Biophysica Acta,* General Subjects, 141, 401-409.

Pan Y. (1995). ^{31}P-^{19}F rotational-echo, double resonance nuclear magnetic resonance experiment on fuoridated hydroxyapatite. *Solid State Nuclear Magn. Reson.*, 5, 263-268.

Panda R.N., Hsieh M.F., Chung R.J., Chin T.S. (2003). FTIR, XRD, SEM and solid state NMR investigations of carbonate-containing hydroxyapatite nano-particles synthesized by hydroxide-gel technique. *J. Phys. Chem. Solids*, 64, 193-199.

Paschalis, E.P., DiCarlo E., Betts F., Sherman P., Mendelsohn R., Boskey A.L. (1996). FTIR microspectroscopic analysis of human osteonal bone. *Calcif. Tissue Int.*, 59, 480-487.

Pasteris J.D., Wopenka B., Freeman J.J., Rogers K., Valsami-Jones E., van der Houten J.A.M., Silva M.J. (2004). Lack of OH in nanocrystalline apatite as a function of degree of atomic order: implications for bone and biomaterials. *Biomaterials*, 25, 229-238.

Penel G., Leroy G., Rey C., Brès E. (1998). MicroRaman spectral study of the PO_4 and CO_3 vibrational modes in synthetic and biological apatites. *Calcif. Tissue Int.*, 63, 475-481.

Pioletti D.P., Takei H., Lin T., Van Landuyt P., Ma Q.J., Kwon S.Y., Sung K.L. (2000). The effects of calcium phosphate cement particles on osteoblast functions. *Biomaterials*, 21,1103-14.

Pors Nielsen S. (2004). The biological role of strontium. *Bone,* 35, 583-588.

Posner A.S., Perloff A, Dioro A.F. (1958). Refinement of hydroxyapatite structure. *Acta Cryst.*, 11, 308-309.

Rey C., Collins B., Goehl T., Dickson R.I., Glimcher M.J. (1989). The carbonate environment in bone mineral. A resolution enhanced Fourier transform infrared spectroscopy study. *Calcif Tissue Int,* 45, 157-164.

Rey C., Shimizu M., Collins B., Glimcher M.J. (1990). Resolution enhanced Fourier transform infrared spectroscopic study of the environment of phosphate ion in the early deposits of a solid phase of calcium phosphate in bone and enamel and their evolution with age : I-investigation in the ν4 PO4 domain. *Calcif. Tissue Int.*, 46, 384-394.

Rey C., Strawich E., Glimcher M.J. (1994). Non-apatitic environments in Ca-P biominerals; implications in reactivity of the mineral phase and its interactions with organic matrix constituents. In D. Allemand, J.P. Cuif (Eds.), *Biomineralizations,* pp. 55-64. Bulletin Océanographique n° special 14 1, Monaco.

Rey C., Kim H.M., Glimcher M.J. (1994) Maturation of poorly crystalline synthetic and biological apatites. In P.W. Brown, B. Constantz (Eds.), *Hydroxyapatite and Related Materials*, pp. 181-187. Boca Raton, CRC Press.

Rey C., Hina A., Tofighi A., Glimcher M.J. (1995). Maturation of poorly crystalline apatites: chemical and structural aspects in vivo and in vitro. *Cells and Mater.,* 5, 345-356.

Rey C., Miquel J.L., Facchini L., Legrand A.P., Glimcher M.J. (1995). On the question of hydroxyl groups in bone mineral. *Bone,* 16, 583-586.

Rey C., Combes C., Drouet C., Sfihi H., Barroug A. (2006). Physico-chemical properties of nanocrystalline apatites: Implications for biominerals and biomaterials. *Materials Science and Engineering C* (in press).

Roberts J.E., Bonar L.C., Griffin R.G., Glimcher M.J. (1992). Characterization of very young mineral phases of bone by solid state ^{31}Phosphorus magic angle sample spinning nuclear magnetic resonance and X-ray diffraction. *Calcified Tissue Int.,* 50, 42-48.

Rodrigues A., Lebugle A. (1998). Influence of ethanol in the precipitation medium on the composition, structure and reactivity of tricalcium phosphate. *Colloid and Surfaces* A, 145, 191-204.

Rogers K., Etok S., Broadhurst A., Scott R. (2005). Enhanced analysis of biomaterials by synchrotron diffraction. *Nuclear Instruments and Methods in Physics Research,* Section A, 548, 123-128.

Rothewel W.P., Waugh J.S, Yesinowsk J.P. (1980). High-resolution variable temperature ^{31}P NMR of solid calcium phosphates. *J. Am. Chem. Soc.,* 102, 2637-2643.

Roufosse A.H., Aue W.P., Roberts J.E., Glimcher M.J., Griffin R.G. (1984). Investigation of the mineral phases of bone by solid state phosphorus-31 magic angle spinning nuclear magnetic resonance. *Biochemistry,* 23, 6115-6120.

Rusu V.M., Ng C.H., Wilke M., Tiersch B., Fratzl P., Peter M.G. (2005). Size-controlled hydroxyapatite nanoparticles as self-organized organic-inorganic composite materials. *Biomaterials,* 26, 5414-5426.

Sarda S., Tofighi A., Hobatho M.C., Lee D., Rey C. (1999). Associations of low temperature apatites ceramics and proteins. *Phosphorus Res. Bull.,* 10, 208-213.

Scherrer P. (1918). Estimation of the size and internal structure of colloidal particles by means of R.overddot.ontgen rays. *Nachr. Ges. Wiss.,* Gottengen, 96-100.

Sfihi H., Rey C. (2002). 1-D and 2-D double heteronuclear magnetic resonance study of the local structure of type B carbonate fluoroapatite. In J. Fraissard, B. Lapina (Eds.) *Magnetic Resonance in Colloid and Interface Science, Nato ASI Series II,* pp. 409-418. Kluwer Academic Publishers.

Shirkhanzadeh M. (1994). X-ray diffraction and Fourier transform infrared analysis of nanophase apatite coatings prepared by electrocrystallization. *NanoStructured Materials,* 4, 677-684.

Sudarsanan K., Young R.A. (1969). Significant precision in crystal structural details: Holly Springs hydroxyapatite. *Acta. Cryst.*, B25, 1534-1543.

Suvorova E.I., Buffat P.A. (1999). Electron diffraction from micro-and nanoparticles of hydroxyapatite. *Journal of Microscopy,* 196, 46-58.

Tadic D., Beckmann F., Schwarz K., Epple M. (2004). A novel method to produce hydroxyapatite objects with interconnecting porosity that avoids sintering. *Biomaterials*, 25, 3335-3340.

Tampieri A., Celotti G., Landi E., Sandri M., Roveri N., Falini G. (2003). Biologically inspired synthesis of bone-like composite: self–assembled collagen fibers/hydroxyapatite nanocrystals. *J Biomed Mater Res,* 67, 618-625.

Termine J.D., Posner A.S. (1966). Infrared analysis of rat bone: age dependency of amorphous and crystalline fractions. *Science,* 153, 1523-1525.

Termine J.D., Eanes E.D., Conn K.M. (1980). Phosphoprotein modulation of apatite crystallization. *Calcified Tissue International,* 31, 247-251.

Thian E.S., Huang J., Vickers M.E., Best S.M., Barber Z.H., Bonfield W. (2006). Silicon-substituted hydroxyapatite (SiHA): a novel calcium phosphate coating for biomedical applications. *J. Mat. Sci*., 41, 709-717.

Tropp J., Blumenthal N.C., Waugh J.S. (1983). Phosphorus NMR study of solid calcium phosphate. *J. Am. Chem. Soc.,* 105, 22-26.

Tsuboi S., Nakagaki H., Ishiguro K., Kondo K., Mukai M., Robinson C., Weatherell J.A. (1994). Magnesium distribution in human bone. *Calcified Tissue International,* 54, 34-37.

Wilson R.M., Elliott J.C., Dowker S.E.P., Rodriguez-Lorenzo L.M. (2005). Rietveld refinements and spectroscopic studies of the structure of Ca-deficient apatite. *Biomaterials*, 26, 1317-1327.

Winand L. (1961). Etude physico-chimique du phosphate tricalcique hydraté et de l'hydroxylapatite. Ann Chim (Paris) 13th series, 6, 951-967.

Wu Y., Glimcher M.J., Rey C., Ackerman J. (1994). A unique protonated phosphate group in bone mineral not present in synthetic calcium phosphates. *J Mol Biol*, 244, 423-435.

Yesinovski J.P., Eckert H. (1987). Hydrogen environments in calcium phosphates: ^{1}H MAS NMR at high spinning speeds. *J. Am. Chem. Soc.,* 109, 6274-6282.

Yesinovski J.P. (1998). Nuclear magnetic resonance spectroscopy of calcium phosphates. In Z. Amjad (Ed.), *Calcium phosphates in biological and industrial systems,* pp. 103-143. Kluwer Academic Publishers.

Yuan H., Li Y., de Bruijn J.D., de Groot K., Zhang X. (2000). Tissue responses of calcium phosphate cement: a study in dogs. *Biomaterials*, 21,1283-1290.

Zahidi E., Lebugle A., Bonel G. (1985). Sur une nouvelle classe de matériaux pour prothèses osseuses ou dentaires. *Bull Soc Chim Fr*, 4, 523-537.

In: Biomaterials Research Advances
Editor: J. B. Kendall, pp. 145-182
ISBN: 978-1-60021-892-7

Chapter 6

STRUCTURE STUDIES OF HYDROXYAPATITE BASED BIOMATERIALS

Th. Leventouri
Department of Physics & Center for Biological and Materials Physics,
Florida Atlantic University, Boca Raton, FL 33431, USA

ABSTRACT

Biological and synthetic hydroxyapatites (HAp) display crystal structure similarities and differences that affect greatly the bioactivity of the synthetic materials. Crystal structure studies from x-ray (XPD) and neutron powder diffraction (NPD) of HAp based biomaterials is discussed in this chapter.

First, a comparison of structural parameters of natural and synthetic apatites from Rietveld refinements of high-resolution NPD patterns as a function of temperature is presented. The natural apatite samples are a carbonate fluorapatite (francolite) and a fluorapatite (harding pegmatite); the synthetic ones are low temperature HAp, and carbonated HAps. Modification of the structural parameters due to the carbonate substitution show a systematic behavior that is consistent with the mechanism of carbonate substitution on the mirror plane of the phosphate tetrahedron.

Then, the effect of silicon substitution on the crystal structure parameters of HAp is discussed from Rietveld refinement analysis of high-resolution NPD patterns as a function of temperature from samples of pure and 0.4 wt % silicon substituted HAp. Small structural changes in the lattice constants, interatomic distances, site occupancies and distortion of the phosphate tetrahedron were found.

In the third part, the structural and magnetic properties of ferrimagnetic bioglass ceramics (FBC) in the system $[0.45(CaO,P_2O_5)(0.52\text{-}x)SiO_2\ xFe_2O_3\ 0.03Na_2O]$, x=0.05, 0.10, 0.15, 0.20, as prepared and after heat treatment in the temperature range 600-1100 °C are assessed. Structure and microstructure of the materials as a function of temperature are studied using x-ray diffraction, scanning electron microscopy, and energy dispersive x-ray spectroscopy. The magnetic properties of FBC are correlated with their bulk and surface structure.

Finally, experimental results of the effect of simulated body fluids (SBF) on the crystal structure and microstructure of FBC are presented.

INTRODUCTION

Hydroxyapatite (HAp), chemical formula $Ca_5(PO_4)_3OH$, is considered the structural template for the mineral phase of bone, dentin, and enamel that comprises ~60 to 70% of the total dry bone weight while the remainder consists of organic materials, such as collagen. The bone mineral is an apatitic calcium phosphate containing carbonate and small amounts of Na, Mg, Si and other trace components. [1,3]

Synthetic HAp is widely used in medicine and dentistry because of its biocompatibility and bioactivity properties. The biological HAps have multiple substitutions and deficiencies at all ionic sites therefore substituted synthetic HAps are studied extensively; a close relationship between substitutions and bioactivity has been demonstrated by *in vitro* and *in vivo* studies [2]. Of particular importance is the B-type carbonate substitution in carbonate HAp (CHAp) where carbonate substitutes for the phosphate ion, because the biological apatites contain 4-6% carbonate by weight that is proven critical in defining their physical properties [2,3].

Minor and trace elements play a major role in the biochemistry of bone, enamel and dentin [2,3,4]. Silicon is one of the trace elements known to be essential in biological processes. As a calcifying agent, it enhances the bone growth rates of bioactive prosthetic materials. The importance of it on bone formation and calcification has been demonstrated through *in vitro* and *in vivo* studies, while the amount of silicon present within active calcification sites is related to "maturity" of the bone mineral [5].

Another category of HAp-based biomaterials is the Ferrimagnetic bioglass ceramics (FBC) that were introduced for hyperthermic treatment of bone cancer [6-8]. They are complex, multiphase, biocompatible and bioactive materials that were fabricated from the original bioglass [9] with the addition of Fe_2O_3 in the system [SiO_2, P_2O_5, CaO]. Certain compositions of glass-ceramics that develop an adherent interface with tissues have been shown to form a mechanically strong bond to bone and are known as "bioactive ceramics" [10-12]. The bioactivity of FBC is related to calcium phosphates and calcium silicates that form as major phases in these compounds; in a physiological environment they partly convert to HAp that is the generally accepted model phase for bone mineral [13].

The magnetic properties of FBC arise from the magnetite [Fe_3O_4] that is produced from the Fe_2O_3 of the starting reacting oxides. FBC have been shown to be bioactive and effective in hyperthermic treatment of animal bone cancer [14-15]. When a FBC material is placed in the region of the tumor and it is subjected to an alternating magnetic field, heat is generated by hysteretic ferrimagnetic loss. The tumor is effectively heated and the temperature locally rises to 42-45 ^{o}C, even if the tumor is deeply seated, because living tissue does not absorb a magnetic field. As a result, the cancerous cells perish while the healthy ones survive.

In this work we mainly focus on the crystal structure and microstructure related properties of hydroxyapatite-based biomaterials that basically determine their behavior as bioactive implanted materials

I. BIOLOGICAL AND SYNTHETIC HYDROXYAPATITES (HAP)

Crystal structure properties of natural and synthetic B-type CHAps have been studied extensively because functions such as solubility, strain, thermal stability, birefringence and bioactivity of the synthetic material, are controlled by their crystallographic structure.

While researchers agree that the planar carbonate ion substitutes for the phosphate tetrahedron in the apatite structure, there is no consensus on the exact arrangement of the substituting planar ion at the tetrahedral phosphate site. Lack of quality single crystals of B-type CAps has kept the problem unresolved in spite of research efforts for more than fifty years. It has become a major issue because the detailed crystallographic structure properties are related to the vacancies created in the lattice by the carbonate substitution, the fractional atomic coordinates and orientation of the carbonate group, which consequently determine the physical properties of the material.

Here we discuss here only the recent works on the carbonate substitution problem starting with the ones that report on different faces of the phosphate tetrahedron as the location of the planar carbonate ion [16-18]. According to Wilson et al. [16] the carbonate ion occupies randomly one of the mirror symmetry related faces of a vacated phosphate ion site with the normal to the plane of the carbonate ion at 30° to the c-crystallographic axis. The structural model they suggest is based on a best Rietveld refinement fit of room temperature x-ray or neutron diffraction patterns of synthetic carbonate HAp (12.5 wt% CO_3 content). In a room temperature x-ray diffraction study of a synthetic, calcium deficient carbonate apatite (4.4 wt% CO_2) Ivanova et al. [17] had proposed two orientations for the carbonate triangle. These triangles share one of their edges, randomly occupy the adjacent faces of a phosphate tetrahedron, and are parallel to the c-axis. Fleet et al. report [18] on the location of B-type carbonate ion using x-ray data from a disordered type A-B carbonate apatite single crystal. According to the authors, the carbonate ion is located close to the sloping faces of the phosphate tetrahedron.

We have investigated [19-21] the carbonate substitution in natural and synthetic B-type carbonate apatites as a function of temperature and carbonate concentration using neutron powder diffraction and the Rietveld analysis method [22]. A systematic behavior of the crystal structure parameters in natural, synthetic, carbonate and non-carbonate apatite is revealed from these experiments. The studied samples are listed in Table I.

TABLE I. Apatite Samples

(a) NATURAL
CFAp: Carbonate-fluorapatite (Francolite of Epirus, Greece) [23]. It contains 4.8 wt% CO_2
CFAp730: The CFAp specimen heat-treated at 730 °C. It contains 3.6 wt% CO_2
FAp: Fluorapatite (Harding pegmatite, Taos County, New Mexico).
(b) SYNTHETIC
CHAp15M: Carbonate hydroxyapatite, $Ca_{5-x}Na_x(PO_4)_{3-x}(CO_3)_xOH$, prepared via hydrolysis of 0.15 molarity solution of $NaHCO_3$. 6.2 wt% CO_2.
CHAp25M: Carbonate hydroxyapatite of the formula $Ca_{5-y}Na_y(PO_4)_{3-y}(CO_3)_yOH$ prepared via hydrolysis of 0.25 molarity solution of $NaHCO_3$. 7.5 wt% CO_2.
HAp: Hydroxyapatite, $Ca_5(PO_4)_3OH$, from SIGMA.

Thermogravimetric analysis (TGA) of the synthetic carbonate apatites has shown a 4.7 wt% loss of water at 530 °C and a loss of 7.5 wt% CO_2 at 840 °C for the sample CHAp25M. A 4.5 wt% loss of water and 6.2 wt% of CO_2 was found for the CHAp15M correspondingly while the carbonate content of the CFAp was 4.8 wt%.

Neutron powder diffraction patterns were measured with the HB-4 high-resolution powder diffractometer, and wavelength λ = 1.0918 Å, at the High Flux Isotope Reactor (HFIR) of the Oak Ridge National Laboratory (ORNL). Data were collected in the angle range 11°-135° as a function of temperature, from room temperature (RT) down to 11 K. The structural parameters were refined using the GSAS program [24]. The crystallographic model [25] used space group P63/m with anisotropic atomic displacement parameters as listed in Table II. The RT patterns were refined first. The scale factor, background, peak profile (pseudo-Voigt), lattice parameters, atomic positions, fractions, and anisotropic displacement parameters were simultaneously refined for all the atoms (except for the deuterium). The positional parameters of the two atoms in the channel were successfully refined. Attempts to refine the temperature parameters of the deuterium atom were not successful, probably due to the low occupancy (0.5) at that site. Although powder samples were measured, the refined temperature parameters of the atoms in the synthetic HAps are comparable with those reported from measurements of single crystal HAps [25]. The weighted residual factors R_{wp} for all patterns, at various temperatures were $0.05 \leq R_{wp} \leq 0.08$ and the reduced χ^2 were $1.1 \leq \chi^2 \leq 1.5$. The RT refined positional parameters, fractions and equivalent isotropic displacement parameters of the atoms in the HAp structure are listed in Tables III, IV, and V; numbers in parentheses show the estimated uncertainties.

TABLE II. Structural parameters of the starting model used to refine the powder diffraction patterns in space group $P6_3/m$, α = 9.41 Å, *c* = 6.88 Å.

Atom	Site Symmetry		χ	y	z	Fraction	Uiso Å^2
Ca1	4*f*	3	0.3333	0.6667	0.0020	1.0	0.002
Ca2	6*h*	*m*	0.2460	0.9915	0.2500	1.0	0.002
P	6*h*	*m*	0.3977	0.3674	0.2500	1.0	0.002
O1	6*h*	*m*	0.3281	0.4819	0.2500	1.0	0.002
O2	6*h*	*m*	0.5862	0.4661	0.2500	1.0	0.002
O3	12*I*	1	0.3436	0.2570	0.0711	1.0	0.002
O(D)	2*α*	$\bar{6}$	0	0	0.1957	0.5	0.002
D	2*α*	$\bar{6}$	0	0	0.0600	0.5	0.004

TABLE III. Refined structural parameters for the synthetic carbonate hydroxyapatite (CHAp 15M) at 295K. α = 9.3958(3) Å, c = 6.8951(4) Å, χ^2 = 1.2, R_{wp} = 6%.

Atom	Site symmetry		*x*	*y*	*z*	Fraction	*Ueq* *100 (Å²)
Ca1	4*f*	3	1/3	2/3	0.0024(13)	0.98	0.87
Ca2	6*h*	*m*	0.245(11)	0.993(1)	¼	0.99	0.94
P	6*h*	*m*	0.3997(10)	0.3727(9)	¼	0.97	0.80
O1	6*h*	*m*	0.3261(7)	0.4819(8)	¼	1.0	0.73
O2	6*h*	*m*	0.5842(8)	0.4644(9)	¼	1.0	1.42
O3	12*i*	1	0.3434(7)	0.2593(7)	0.0725(6)	1.0	2.51
O(D)	2*α*	$\bar{6}$	0	0	0.209(3)	0.5	3.55
D	2*α*	$\bar{6}$	0	0	0.098(3)	0.5	4.00

TABLE IV. Refined structural parameters for the synthetic carbonate Hydroxyapatite (CHAp 25M) at 295K. α = 9.3927(3) Å, c = 6.8985(4) Å, χ^2 = 1.1, R_{wp} = 7%

Atom	Site symmetry		*x*	*y*	*z*	Fraction	100**U*eq (Å²)
Ca1	4*f*	3	1/3	2/3	0.003(1)	0.98(1)	0.50
Ca2	6*h*	*m*	0.245(1)	0.991(1)	1/4	0.98(1)	0.95
P	6*h*	*m*	0.400(1)	0.374(1)	1/4	0.94(1)	0.90
O1	6*h*	*m*	0.3268(7)	0.4840(8)	1/4	1.0	0.85
O2	6*h*	*m*	0.5828(8)	0.4637(8)	1/4	1.0	1.96
O3	12*i*	1	0.3428(7)	0.2580(7)	0.0726(6)	1.0	2.72
O(D)	2a	$\bar{6}$	0	0	0.199(2)	0.5	1
D	2a	$\bar{6}$	0	0	0.080(3)	0.5	4

TABLE V. Refined structural parameters for the commercial hydroxyapatite (HAp) at 295K. α = 9.4183(3) Å, c = 6.8843(3) Å, χ^2 = 1.4, R_{wp} = 6%.

Atom	Site symmetry		*x*	*y*	*z*	Fraction	100*U_{eq} (Å^2)
Ca1	4*f*	3	1/3	2/3	0.002(1)	0.97(1)	1.51
Ca2	6*h*	*m*	0.2462(8)	0.9925(9)	1/4	0.95(1)	0.58
P	6*h*	*m*	0.4013(13)	0.3724(8)	1/4	0.99(1)	0.57
O1	6*h*	*m*	0.3274(6)	0.4827(8)	1/4	1.0	0.70
O2	6*h*	*m*	0.5882(6)	0.4645(8)	1/4	1.0	1.18
O3	12*i*	1	0.3444(6)	0.2576(6)	0.0699(5)	1.0	1.96
O(D)	2a	$\bar{6}$	0	0	0.207(2)	0.5	1.40
D	2a	$\bar{6}$	0	0	0.139(3)	0.5	5

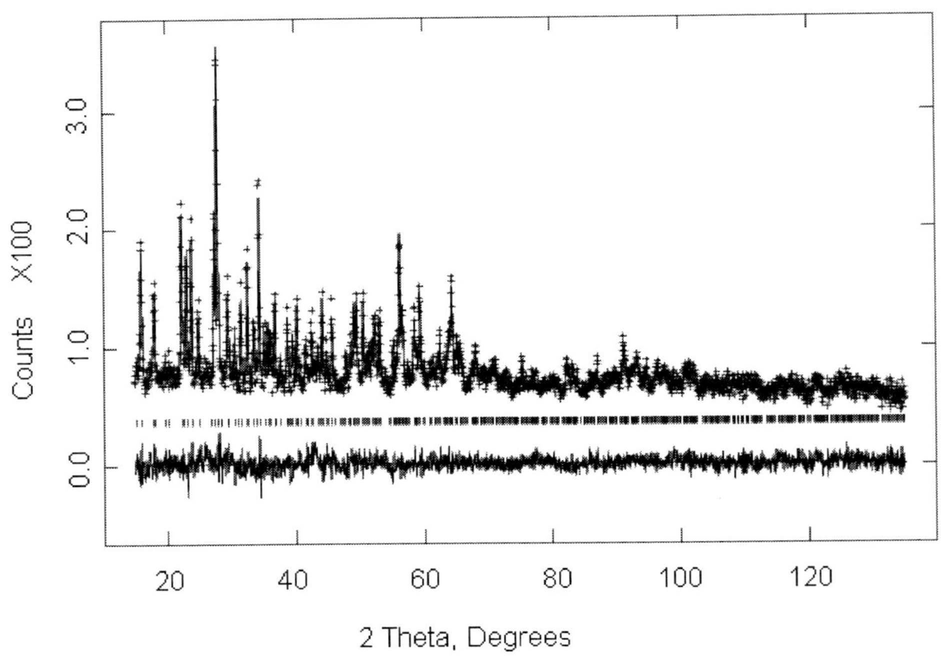

Figure I.1. Rietveld refinement fit of the RT neutron powder diffraction pattern (λ = 1.0918 Å) from the synthetic HAp (sample CHAp25M). The crosses show the observed data (experimental pattern), and the solid line is the calculated diffraction pattern. The vertical black lines mark the positions of the calculated Bragg peaks and the lower trace is the difference between the observed and calculated patterns.

Figure I.1 shows an example of a Rietveld refinement fit from the RT neutron powder diffraction pattern ($\lambda = 1.0918$ Å) of the synthetic HAp, sample CHAp25M.

The refined fractions of the P atom as a function of the carbonate content are plotted in Figure I.2 for synthetic (circles) and natural (squares) HAps. Notice that from 0.99 in the non-carbonate synthetic HAp, or 1 for the natural HAp, which correspond to a site fully occupied by P, the site occupancy decreases to 0.93 for the sample CHAp25M. Reduction of the occupancies from 1 to 0.93 with the carbonate content is a strong indication of carbonate substitution at the phosphate site.

The isotropic U_{eq} for the P atom also increases with the carbonate concentration from 0.0057 Å^2 for the non-carbonate HAp (Table V) to 0.0080 Å^2 for the CHAp15M (Table III) and to 0.0090 for the sample CHAp25M (Table IV). Since U_{eq} is a measure of the atomic displacement, such increase also suggests that the P site is where the carbonate substitution occurs. The U_{eq} for the atoms O1, O2, and O3 increase with the carbonate concentration indicating the disturbance of the PO_4 tetrahedron as the planar carbonate ion enters the apatite structure.

Several structural parameters of the pure HAp and the synthetic carbonate HAps show a systematic behavior that is reflected on the similarities of the physical properties of carbonate apatites. Their lattice constants confirm the expected contraction of the α-lattice parameter and slight expansion of the c-parameter with the increase of carbonate content [2, 3, 26, 27]Table I.3 shows the temperature dependence of the lattice constants of the synthetic HAp samples listed in Table I.

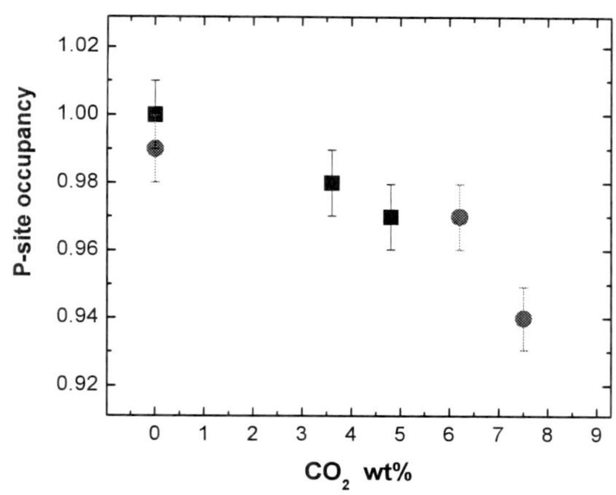

Figure I.2. The refined occupancies of the phosphorus atom at the tetrahedral site as a function of the measured CO_2 concentration. Circles are used for the synthetic and squares for the natural samples listed in Table I. Error bars are drawn as a 1 % of the refined occupancies.

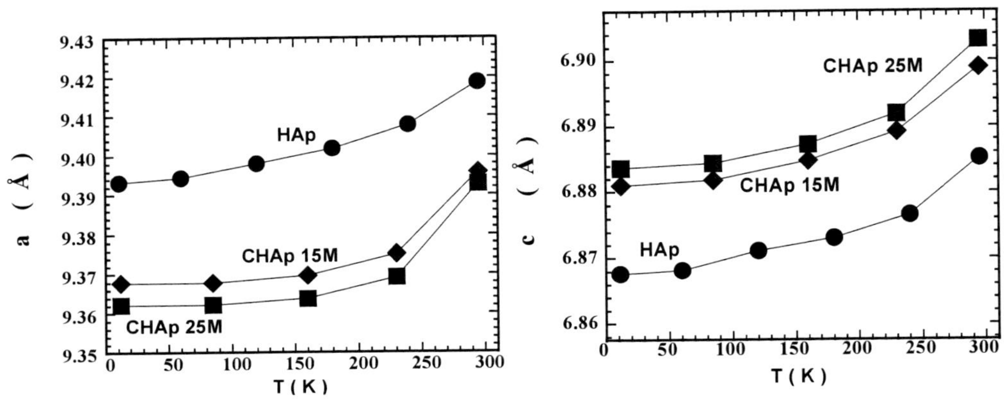

Figure I.3. Temperature dependence of the α and c-lattice constants of the commercial hydroxyapatite (sample HAp) and the synthetic carbonate hydroxyapatites (samples CHAp15M and CHAp25M). Error bars do not show in this scale.

The bond lengths of the tetrahedral site as a function of temperature are plotted in Figure I.4 for all the HAp samples of Table I. While the bond lengths between the phosphorus and the oxygen atoms remain invariable with the temperature, within errors, there is a clear distinction between the P-O3 bond lengths of all the samples and the P-O1, P-O2 of the two carbonate apatite specimens. The upper area of the plot, between 1.52 and 1.55 Å includes, within error, all the P-O3 values of the CHAp15M and CHAp25M, as well as the P-O1, P-O2 and P-O3 for the HAp. The lower part between 1.48 and 1.52 Å contains the P-O1 and P-O2 bonds of the carbonate apatites only. They display systematically distorted P-O1 and P-O2 bonds compared to the undistorted value of 1.54 Å. Such disturbance indicates that the mirror plane is the position of the carbonate since the P-O1 and P-O2 bonds are disturbed by the carbonate substitution, whereas the P-O3 bonds are not. Such systematic reduction can be explained by the carbonate with its shorter bonds (1.13 Å) entering the apatite structure on the mirror plane of the phosphate tetrahedron. Others have reported similar decrease of the bond lengths of carbonate apatites. For instance, V. Perdikatsis [23] has reported ~3% reductions of the P-O1 and P-O2 bonds from Rietveld analysis of x-ray diffraction data of CFAp (francolite). Wilson et al. have found [16] a reduction of ~6% of the P-O2 bond for human enamel carbonate hydroxyapatite. Recent work on natural CFAp reports ~ 3 % and 4 % decrease in P-O1 and P-O2 bond lengths correspondingly from powder neutron diffraction experiments [19, 20, 21].

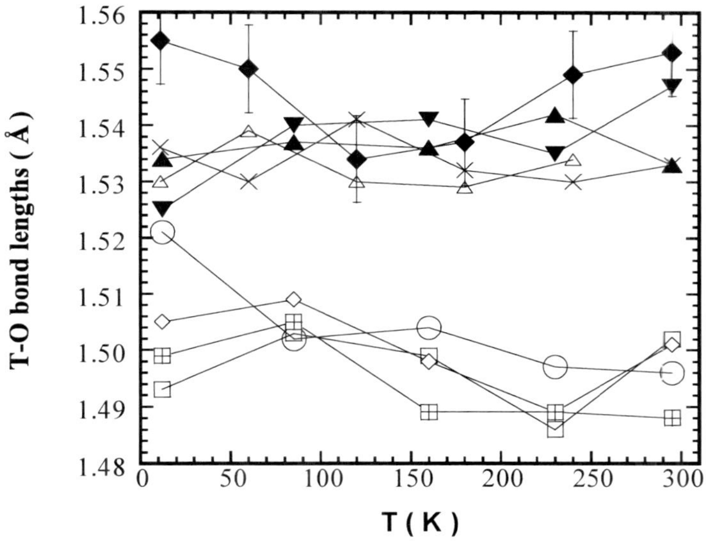

Figure I.4. Interatomic distances of the tetrahedral site (T) as a function of temperature for synthetic carbonate hydroxyapatites in comparison with the bond lengths of the non-carbonate hydroxyapatite. Open symbols mark the interatomic distances P-O1: Circles for the sample CHAp15M, diamonds for CHAp25M and triangles for HAp. Open symbols also show the P–O2 bonds: squares the sample CHAp15M, divided squares the CHAp25M, and × the HAp. Filled symbols mark the P-O3 bonds of the three samples: Upward triangles the CHAp15M, downward triangles the CHAp25M and diamonds the HAp. Error bars are drawn only for the P-O3 bonds of the non-carbonate hydroxyapatite (HAp) as a percentage of their refined values.

The distortion index (DI) [28] of the phosphate tetrahedron was calculated from the equation:

$$DI = \frac{\left(\sum_{i=1}^{4}\left|TO_i - TO_m\right|\right)}{4TO_m}$$

In this equation TO_i denotes the interatomic distances P-O, the i subscript refers to the individual value, and m to the mean value. The distortions, calculated from the refined T-O bonds were found between 1.5 and 1.8% for the CAps compared to 0.3% for the non carbonate apatite (FAp) [19].

Figure I.5 shows a c-axis projection of three unit cells generated from the RT refined NPD pattern of the sample CHAp15M with anisotropic temperature parameters. In this figure, the probability density function ellipsoids are plotted for the atoms. The size and orientation of these ellipsoids depend on the anisotropic temperature factors U_{ij} of the individual atom at its specific site.

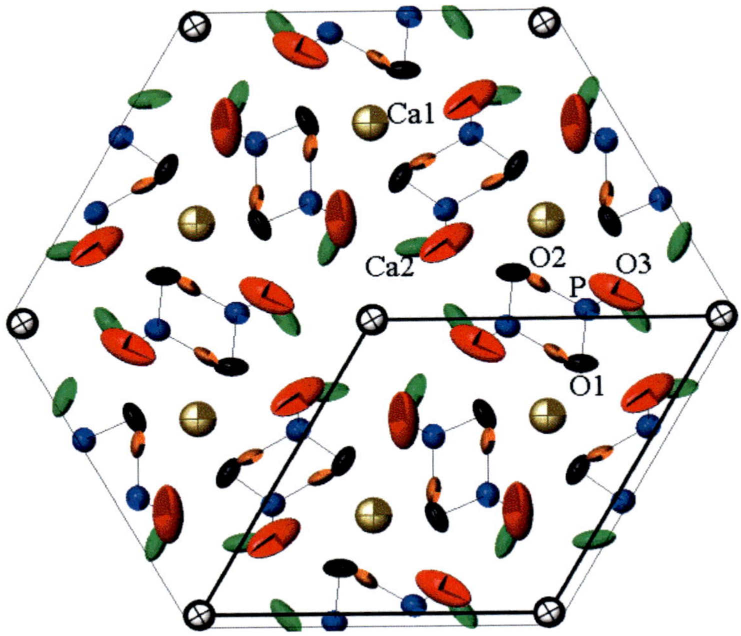

Figure I.5. A c-axis projection for 3 unit cells from the RT refined NPD pattern of the carbonate hydroxyapatite sample CHAp15M with anisotropic temperature factors. Only the P-O bonds of the phosphate tetrahedron are plotted. Blue is used for the P atoms, black for O1, orange for O2, and red for O3. Gold marks the Ca1 atoms and green the Ca2. The hydroxyl is shown at the channel position (c-crystallographic axis).

Detailed studies of the refined crystal structure parameters of several carbonated and carbonate-free samples as a function of temperature have revealed a great deal of lattice distortion on the mirror plane of the phosphate tetrahedron.

The Atomic Displacement Parameters (ADPs) U(T) as a function of temperature reveal the extent of static versus dynamic disorder. If the zero-point vibrations are neglected or assumed to be the same for similar atoms, the zero temperature intercept of U(T) is proportional to the static disorder, i.e., the positional disorder. In that contest, Figure I.6 shows in blue bars the equivalent ADPs U_{eq} of the atoms of the phosphate tetrahedron of a synthetic HAp from RT down to 11K, as calculated from the refined ADPs U_{ij} and extrapolated at 0 K. These U_{eq} are compared with the ones for the synthetic CHAp15M (6.2 wt % carbonate green bars), and the CHAp25M (7.5 wt % carbonate marked with red bars). As the U_{eq} extrapolated at 0 K provide a qualitative measure of the static contribution to the atomic displacement of the corresponding atom in a crystal, they provide insight about the carbonate substitution. Notice that, the atomic displacement for the atom O1 increases up to a factor of ~8 with the increase of the carbonate content. For the atom O2 the atomic displacement increases up to a factor of ~4, while it goes up to a factor of 26 in the case of the P atom. The two O3 atoms, above and below the mirror plane, exhibit large displacement even in the pure HAp due to the low occupancy in hydroxyl at the channel site, while it increases up to a factor of ~2 with the carbonate substitution. Such increase is expected for the O3 atoms because of the created vacancy when CO_3 substitutes for PO_4.

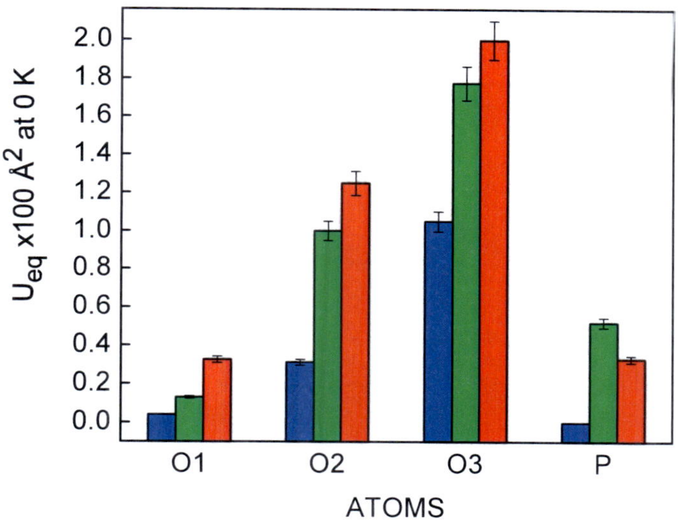

Figure I.6. The equivalent atomic displacement parameters U_{eq} of the atoms at the phosphate tetrahedron as calculated and extrapolated at 0 K from the refined anisotropic temperature parameters U_{ij} of a synthetic HAp shown in blue bars. They are compared with the corresponding U_{eq} of a synthetic carbonate HAp (CHAp15M) containing 6.2 wt % carbonate (green bars), and a synthetic carbonate HAp (CHAp25M) containing 7.5 wt % carbonate (red bars). These extrapolated values of U_{eq} provide a qualitative measure of the static contribution to the atomic displacement parameters of the corresponding atoms.

Studies of the refined ADPs of the carbonate fluorapatite of Table I as a function of temperature in the range 10K to 295K that are listed in Table VI have provided similar type of disorder in the phosphate tetrahedron. The temperature dependence of the ADPs indicates enhanced static disorder associated with the F, O3, and tetrahedral sites. The symmetry constraints for the atomic sites in $P6_3/m$ apatite are such that for the Ca1 and F sites, the U_{ij} components are $U_{11} = U_{22}$, U_{33}, U_{12}, $U_{13} = U_{23} = 0$; for the Ca2, P, O1, and O2 the U_{ij} components are U_{11}, U_{22}, U_{33}, U_{12}, $U_{13} = U_{23} = 0$; and for O3 the U_{ij} components are U_{11}, U_{22}, U_{33}, U_{12}, U_{13}, and U_{23}. It is only for O3 that the probability density function (PDF) ellipsoid is allowed by symmetry to take on any orientation.

The anisotropic equiprobability ellipsoids describing the mean-square atomic displacements at the tetrahedral site are plotted in Figure I.7 using results from a RT refined NPD pattern of francolite. When carbon leaves the structure the whole tetrahedron appears as if it moves towards the channel (z-axis).

TABLE VI. Refined anisotropic displacement parameters (Å^2) of francolite (heated @530°C) in the temperature range 10 to 295 K.

ATOM	100*Uij	10 K	50 K	100 K	150 K	200 K	250 K	295 K
Ca1	U_{11}	0.6(1)	0.7(1)	0.7(1)	1.0(1)	1.2(2)	1.1(2)	1.3(1)
	U_{33}	0.0(1)	0.3(2)	-0.1(2)	0.1(2)	0.2(2)	0.6(2)	0.5(2)
	U_{12}	0.3(1)	0.4(1)	0.4(1)	0.5(1)	0.6(1)	0.6(1)	0.6(1)
Ca2	U_{11}	1.1(1)	1.2(2)	1.1(2)	1.3(2)	1.2(2)	1.5(2)	1.7(2)
	U_{22}	0.6(2)	0.5(2)	0.6(2)	0.8(2)	0.6(2)	1.1(2)	1.0(2)
	U_{33}	0.3(1)	0.3(2)	0.3(2)	0.4(2)	0.4(2)	0.5(2)	1.0(2)
	U_{12}	0.2(1)	0.3(2)	0.0(2)	0.3(1)	0.1(2)	0.5(2)	0.6(2)
P	U_{11}	0.4(1)	0.4(2)	0.5(2)	0.5(2)	0.1(2)	0.7(2)	0.9(2)
	U_{22}	1.5(2)	1.2(2)	1.5(2)	1.6(2)	1.5(2)	1.6(2)	1.9(2)
	U_{33}	1.9(1)	2.0(2)	2.3(2)	2.3(2)	2.0(2)	2.5(2)	2.7(2)
	U_{12}	0.5(1)	0.4(2)	0.6(2)	0.7(2)	0.2(2)	0.9(2)	0.8(2)
O1	U_{11}	1.0(1)	1.0(2)	1.1(2)	1.2(2)	1.4(2)	2.0(2)	2.0(2)
	U_{22}	0.9(2)	0.7(2)	1.0(2)	1.2(2)	1.1(2)	1.1(2)	1.4(2)
	U_{33}	0.7(1)	0.8(2)	0.5(2)	0.7(2)	0.9(2)	1.3(2)	1.2(2)
	U_{12}	0.8(1)	0.7(2)	0.8(2)	0.9(2)	0.9(2)	1.4(2)	1.2(2)
O2	U_{11}	0.6(1)	0.9(2)	0.6(2)	0.4(2)	0.9(2)	0.6(2)	0.9(2)
	U_{22}	0.5(1)	0.7(2)	0.9(2)	0.8(2)	1.1(2)	0.9(2)	1.2(2)
	U_{33}	1.0(1)	0.9(2)	1.4(2)	1.7(2)	1.8(2)	2.0(2)	2.2(2)
	U_{12}	0.5(1)	0.7(2)	0.7(2)	0.5(2)	0.8(2)	0.5(2)	0.8(1)
O3	U_{11}	2.8(1)	2.7(2)	3.1(2)	3.0(2)	3.1(2)	3.4(2)	3.9(2)
	U_{22}	1.3(1)	1.3(1)	1.2(1)	1.6(1)	1.8(1)	2.0(1)	2.1(1)
	U_{33}	0.7(1)	0.7(1)	1.0(1)	1.0(1)	1.1(1)	1.0(1)	1.3(1)
	U_{12}	1.5(1)	1.4(1)	1.5(1)	1.6(1)	1.9(1)	1.9(2)	2.1(1)
	U_{13}	-0.6(1)	-0.8(1)	-0.9(1)	-0.8(1)	-1.0(1)	-1.0(1)	-1.1(1)
	U_{23}	-0.3(1)	-0.4(1)	-0.4(1)	-0.5(1)	-0.5(1)	-0.6(1)	-0.6(1)
F	U_{11}	1.2(2)	1.1(2)	1.0(2)	1.2(2)	1.4(2)	1.4(2)	1.7(2)
	U_{33}	3.2(3)	3.9(5)	4.1(5)	5.0(5)	4.5(5)	5.4(6)	5.7(5)
	U_{12}	0.6(1)	0.5(1)	0.5(1)	0.6(1)	0.7(1)	0.7(1)	0.8(1)

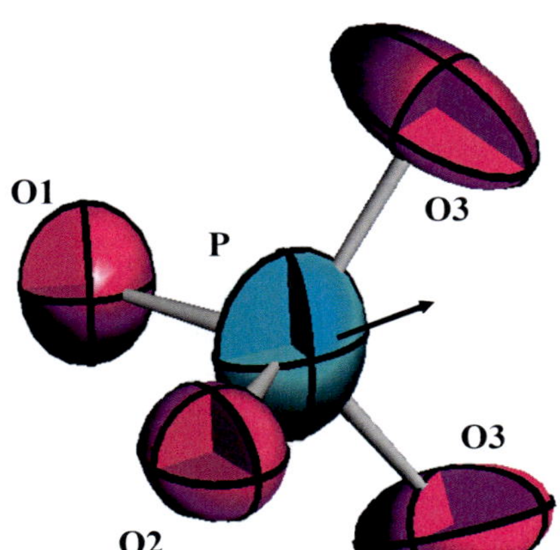

Figure I.7. Equiprobability ellipsoids (drawn at 99% probability) for the anisotropic displacements of the carbonate fluorapatite tetrahedral site atoms.

The RT equivalent isotropic displacement parameters of the atoms in the francolite structure (sample CFAp530) are plotted in Figure I.8. These parameters are compared with the ones for a partially (CFAp730) or completely de-carbonated specimen (CFAp840) and with a FAp (Harding pegmatite). These displacements demonstrate the distortion of the phosphate tetrahedron in CFApsatomic positions of O1, O2 and P in CFAps.

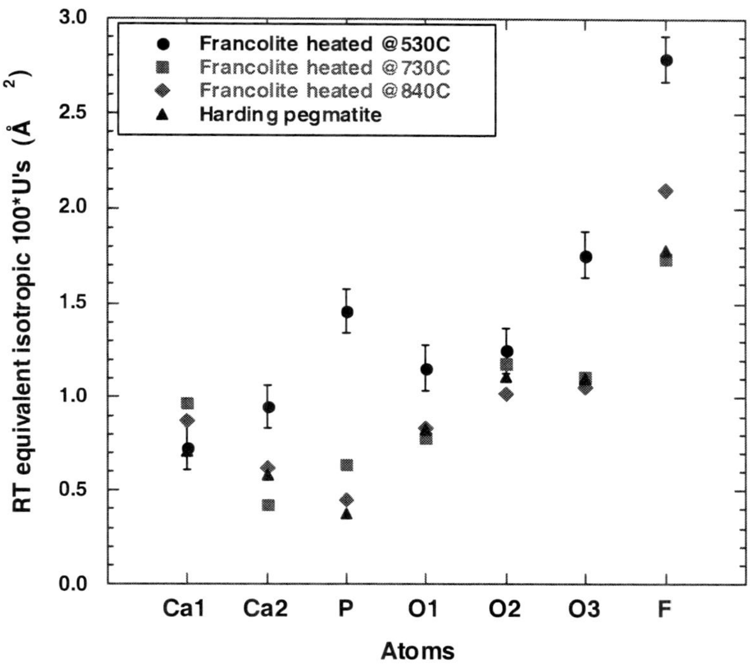

Figure I.8. Room temperature equivalent isotropic displacement parameters of the atoms in the francolite samples with carbonate (F530) and with partially (F730) or completely (F840) removed carbonate, along with the fluorapatite specimen.

The results of the work on synthetic carbonate hydroxyapatites (CHAps) are consistent with the mechanism of carbonate substitution on the mirror plane of the phosphate tetrahedron, as it was introduced for the natural carbonate fluorapatite (CFAp) [21]. Comparison of the results from both types of B-type CAps has shown that the tetrahedral bond lengths P-O1 and P-O2 decrease by 3-4% in all carbonate apatites. The ADPs of the phosphate tetrahedron are greater in the carbonate than in the non-carbonate apatites. Finally, the atomic positional disorder of the T-site (P/C site) is greater in the CFAp than in the CHAps, while the opposite happens at the O3 sites.

II. Silicon Substituted Hydroxyapatite

Silicon substituted hydroxyapatite (SiHAp) has been prepared by using various processing techniques [29-31]. Gibson et al. [29, 30] had prepared a single phase of SiHAp via an aqueous precipitation method. They confirmed the incorporation of 0.4 wt % Si into the HAp structure with replacement of the phosphorus ion by silicon using chemical analysis methods. Rietveld refinement of room-temperature powder x-ray diffraction patterns had shown a small effect of the silicon substitution on the HAp crystal structure.

We had prepared pure and 0.4 wt % silicon substituted HAp following an aqueous precipitation method [30]. The samples of pure and silicon substituted hydroxyapatite were prepared via an aqueous precipitation reaction of $Ca(OH)_2$ and H_3PO_4 with the appropriate amount of silicon acetate [$Si(CH_3COO)_4$]. The as prepared hydroxyapatite (sample HAp), as well as the silicon substituted (sample SiHAp) were heat treated at 1100 °C in order to have their crystallinity improved. The two types of heat-treated samples were named HApS and SiHApS accordingly.

High-resolution neutron powder diffraction (NPD) was used to study the effect of the silicon substitution on the crystal structure parameters of the HAp structure as a function of temperature from RT down to 20 K [32]. The scattering length of the oxygen atom is large enough so that neutron diffraction can detect the expected small changes in the phosphate group. Moreover, hydrogen has a large enough scattering length to provide information on modifications of the hydroxyl ions on the z-crystallographic axis (channel positions).

All the samples were deuturated prior to measurements so that the hydrogen would be exchanged by deuterium which has a smaller incoherent scattering length than hydrogen resulting to a decrease of the background of the diffraction pattern. The four types of studied samples are listed in Table I.

X-ray powder diffraction measurements were performed for phase identification and phase purity with the use of a SIEMENS D5000 diffractometer equipped with a diffracted beam monochromator and CuKα wavelength. A single hydroxyapatite phase (card # 9-432) was identified in all specimens by using the ICDD data bank. Figure II.1(a) shows part of the diffraction patterns from the as-prepared pure (sample HAp) and the Si-substituted hydroxyapatite (sample SiHAp). Figure II.1(b) displays the diffraction patterns of the same samples after they were heat-treated at 1100°C. Comparison of these two figures reveals a considerable improvement of the crystallinity and microstrain in the heat-treated material, as it is shown by the diffraction peaks profile and decrease of the full width at half maximum (FWHM). No secondary phases were identified within the x-rays detection limits of ~2%.

High-resolution neutron powder diffraction data were collected at the High Flux Isotope Reactor (HFIR) of the ORNL with wavelength $\lambda = 1.0913$ Å, scattering angles between 11° and 135°, from room temperature down to 20K. The structural parameters were refined using the program general structure analysis system (GSAS) [24] with isotropic temperature parameters. The values of the weighted residuals R_{wp} were $0.05 \le R_{wp} \le 0.08$ with reduced χ^2 $1.00 \le \chi^2 \le 1.70$, while the R_{Bragg} values were $0.05 \le R_{Bragg} \le 0.11$ for all the diffraction patterns, at various temperatures, and for up to 32 simultaneously varied parameters. The RT refined positional parameters, fractions, and equivalent isotropic displacement parameters for the heat-treated samples HApS and SiHApS are listed in Tables II and III correspondingly.

TABLE I. Hydroxyapatite and Silicon-substituted Hydroxyapatite Samples

(a) As-prepared HAp: Hydroxyapatite SiHAp: 0.4wt % Silicon-substituted Hydroxyapatite
(b) Heat-treated at 1100 °C for 8 h HApS: Hydroxyapatite SiHApS: Silicon-substituted Hydroxyapatite.

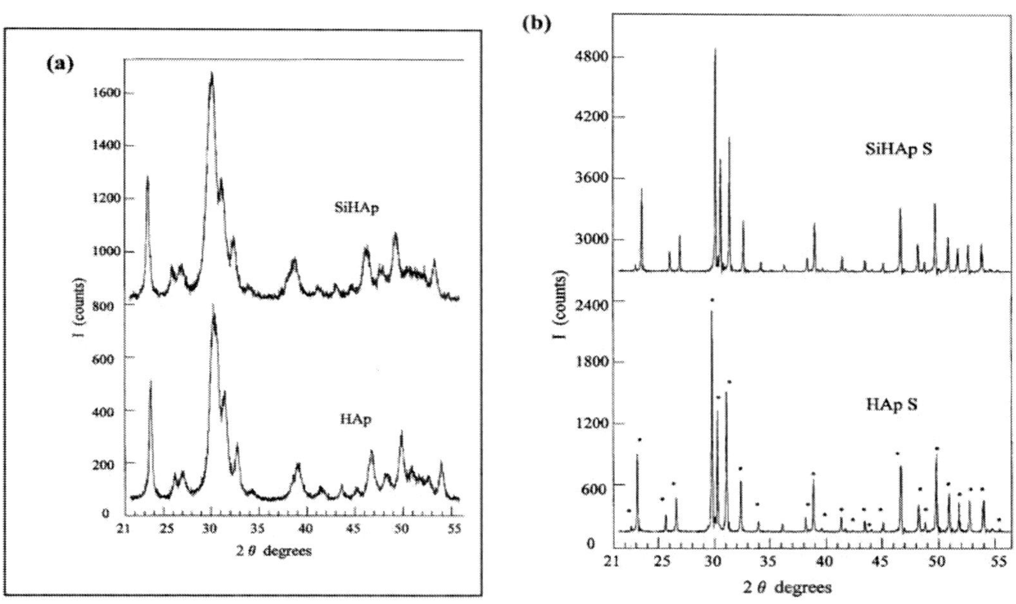

Figure II.1. Part of the x-ray diffraction patterns from the hydroxyapatite (HAp), and silicon substituted hydroxyapatite (SiHAp): as-prepared samples **(a)** and after heat treatment **(b)**. Dots mark the peaks of the hydroxyapatite phase as identified by using the ICDD data bank.

TABLE II. RT refined positional parameters, fractions, (site occupancies) and equivalent isotropic displacement parameters for the heat-treated HApS. The agreement of the fit factors R_{wp} and χ^2 are also listed for each sample.

Heat-treated Hydroxyapatite

HApS

α = 9.4162(1) Å, c = 6.8791(1) Å, V = 528.21(1) Å^3

χ^2 = 1.7 R_{wp} = 7%

Atom	Site symmetry		*x*	*y*	*z*	Fraction	100 U_{eq} Å^2
Ca1	4*f*	3	1/3	2/3	0.0015(7)	0.98(2)	0.76(9)
Ca2	6*h*	*m*	0.2454(5)	0.9928(5)	1/4	0.96(2)	0.52(6)
P	6*h*	*m*	0.3991(5)	0.3688(4)	1/4	1.0	0.48(6)
O1	6*h*	*m*	0.3272(4)	0.4825(4)	1/4	1.0	0.62(5)
O2	6*h*	*m*	0.5883(4)	0.4657(4)	1/4	1.0	1.05(7)
O3	12*i*	1	0.3435(3)	0.2584(3)	0.0705(3)	1.0	1.32(4)
O(D)	2*a*	$\bar{6}$	0	0	0.206(2)	0.30(1)	1.0
D	2*a*	$\bar{6}$	0	0	0.168(3)	0.30(1)	4.0

TABLE III. RT refined positional parameters, fractions, (site occupancies) and equivalent isotropic displacement parameters for the heat-treated SiHApS. The agreement of the fit factors R_{wp} and χ^2 are also listed for each sample.

Heat-treated, Silicon substituted Hydroxyapatite SiHApS

α=9.4157(2) Å, c= 6.8858(1) Å, V=528.68(2) Å^3

χ^2 = 1.4 R_{wp} = 7%

Atom	Site symmetry		*x*	*y*	*z*	Fraction	100 U_{eq} Å^2
Ca1	4*f*	3	1/3	2/3	0.002(1)	0.98(1)	0.8(1)
Ca2	6*h*	*m*	0.2466(7)	0.991(1)	1/4	0.99(1)	0.54(1)
P	6*h*	*m*	0.3992(7)	0.3694(6)	1/4	1.0	0.41(8)
O1	6*h*	*m*	0.3266(6)	0.4823(6)	1/4	1.0	0.77(9)
O2	6*h*	*m*	0.5884(6)	0.4662(6)	1/4	1.0	0.9(1)
O3	12*i*	1	0.3418(5)	0.2568(5)	0.0726(4)	1.0	1.50(7)
O(D)	2*a*	$\bar{6}$	0	0	0.198(9)	0.26(1)	1
D	2*a*	$\bar{6}$	0	0	0.19(1)	0.26(1)	3

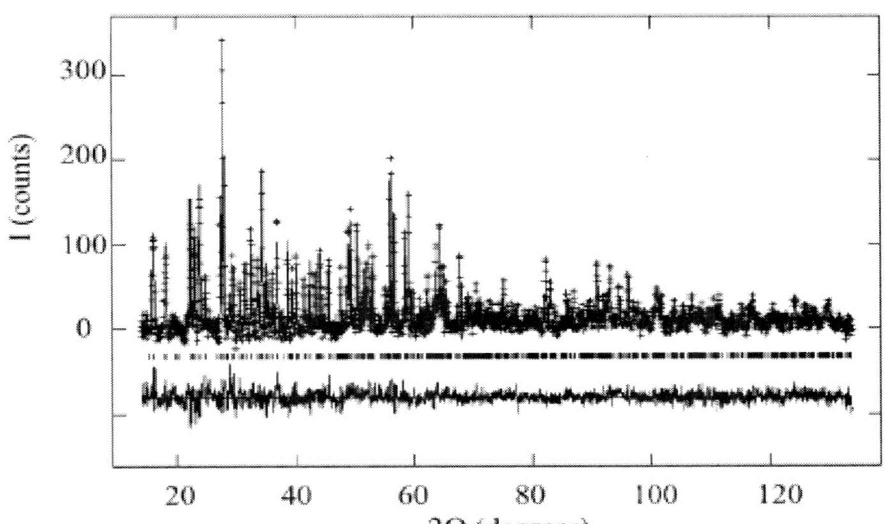

Figure II.2. Rietveld refinement fit of a room-temperature neutron powder diffraction pattern (λ=1.0913 Å) from the heat-treated, silicon substituted hydroxyapatite (sample SiHApS). Crosses mark the experimental pattern and the solid line is the calculated diffraction pattern. The lower trace is the difference between observed and calculated patterns, and the vertical lines mark the positions of the calculated Bragg peaks.

A typical refinement fit between the observed and calculated powder diffraction patterns from the RT data of the sample SiHApS is illustrated in Figure II.2.

The lattice constants as a function of temperature for all the samples are plotted in Figure II.3. Comparison of the RT results for the heat-treated samples of Table II shows that when silicon substitutes for P, the α-lattice constant contracts by ~0.005% while the c-lattice constant expands by ~0.1%. Expansion along the z-axis is favored by the low hydroxyl site occupancy in the channels (30% for the sample HApS), which becomes even lower in the silicon-substituted specimen (26% for the SiHApS).

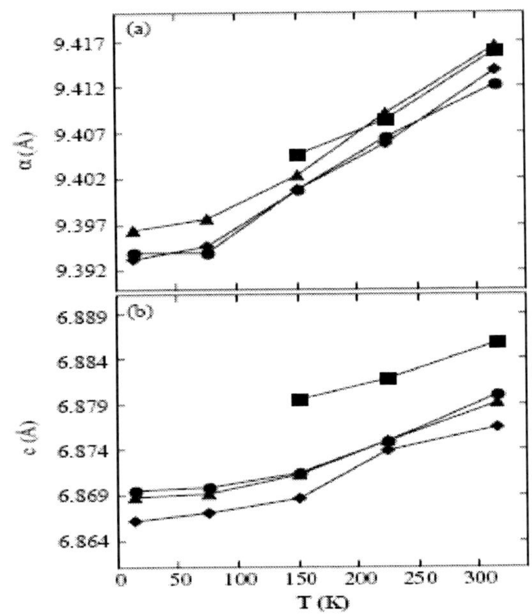

Figure II.3. Temperature dependence of the lattice parameters of the four samples: **(a)** The α lattice constants, and **(b)** the c lattice constants. Diamond symbols are used for the as prepared hydroxyapatite (sample HAp), circles for the as prepared silicon-substituted hydroxyapatite (sample SiHAp), triangles for the heat-treated hydroxyapatite (sample HApS) and squares for the heat-treated silicon-substituted (sample SiHApS). Error bars are smaller than the plot symbols. Lines are guides to the eye.

The unit cell volume expands by a small percentage (~0.09%), as expected from the size proximity of the phosphorus and silicon atoms, as well as from the small percentage of the silicon substitution (0.4 wt%).

The refined values of the interatomic P-O distances of the phosphate tetrahedron are plotted as a function of temperature in Figure II.4 for all the samples. Figure II.4a shows the T-O1 bond lengths, II.4b the T-O2 and II.4c the T-O3 bond lengths. They all show no temperature dependence, within error, while a small effect of the silicon substitution on the T-O1 bond lengths almost vanishes after heat-treating the samples.

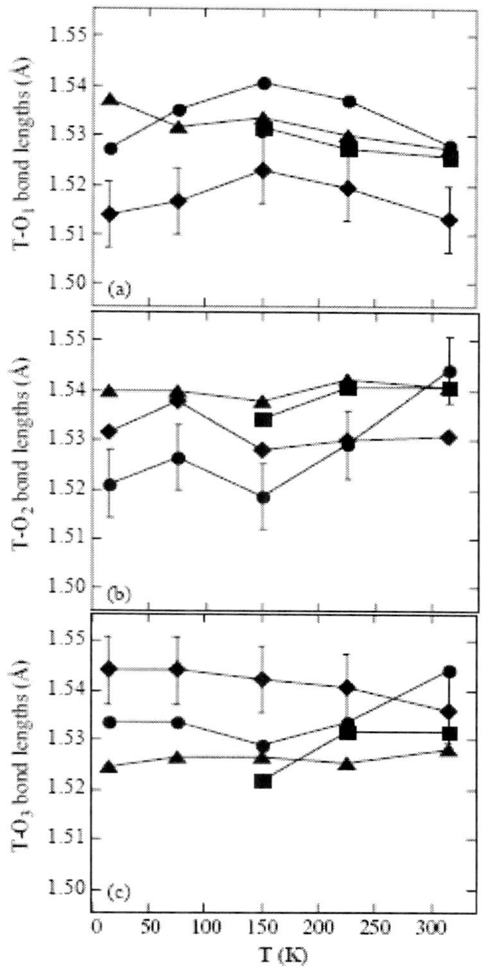

Figure II.4. The interatomic distances at the tetrahedral site (T) as a function of temperature for all the specimens: Part **(a)** shows the T-O1 bond lengths, **(b)** the T-O2 and **(c)** the T-O3. Diamond symbols are used for the as prepared hydroxyapatite (sample HAp), circles for the as prepared silicon-substituted hydroxyapatite (sample SiHAp), triangles for the heat-treated hydroxyapatite (sample HApS) and squares for the heat-treated silicon-substituted (sample SiHApS). Error bars are plotted for one sample in each part of the Figure for clarity. Lines are guides to the eye.

A measure of the effect of the silicon substitution on the hydroxyapatite structure can be found by calculating the angle distortion index of the phosphate tetrahedron [33]from:

$$DI = \frac{\left(\sum_{i=1}^{6} \left| OTO_i - OTO_m \right| \right)}{6(OTO_m)}$$

In this equation OTO_i denotes each one of the six angles between the P/Si and the four O atoms of the phosphate tetrahedron and OTO_m their average. The RT refined values of these angles for the heat-treated samples vary from 106.3° (angle O_3TO_3, sample SiHApS) to 111.8°

(angle O_1TO_2, sample HApS) compared to ~109.5° for an ideal tetrahedron. An evaluation of the DI values as calculated from the refined bond angles of the heat-treated specimens shows that the silicon substitution results to an increase of the angle distortion only by 0.2%. Such small distortion of the phosphate tetrahedron is in accordance with the P and Si atomic size proximity as well as with the small percentage of the substitution.

According to Table II the hydroxyl site occupancy (fraction) decreases from 0.30 in the HApS to 0.26 in the SiHApS. Notice that while in both samples the occupancies are much lower than the starting value of 50%, there is an additional reduction of ~13% in the silicon substituted sample. Such decrease possibly indicates a compensation for the extra negative charge that the $(SiO_4)^{-4}$ ions introduce in the HAp structure when replacing the $(PO_4)^{-3}$ ions. It also includes a compensation for the reduction of the calcium occupancies as listed in Table II.

Finally, the results of the refinements of the neutron diffraction patterns have shown that the heat treatment produces a material of higher density, as expected. The density of the substituted HAp increases from 3.032 gr/cm^3 in the sample SiHAp to 3.089 gr/cm^3 in the heat-treated silicon substituted hydroxyapatite (sample SiHApS).

III. HAp-based Magnetic Biomaterials: Ferrimagnetic Bioglass Ceramics (FBC)

Most research efforts focus on developing magnetic bioceramics for clinical applications by using a variety of compositions and processing methods. However, it is recognized that fundamental studies of physical properties are needed to obtain long-term reliable biomaterials [34-38].

A systematic study of the correlations between processing parameters, crystal structure, microstructure and magnetic properties of FBC is discussed here.

A standard melting processing method was used to prepare four series of samples in the system [$0.45(CaO,P_2O_5)$ $(0.52-x)SiO_2$ xFe_2O_3 $0.03Na_2O$]. The Na_2O was added to reduce viscosity and improve the reactivity of the melting mixture [39,35]. The molar concentrations in Fe_2O_3 were $x = 0.05, 0.10, 0.15, 0.20$. The Ca/P ratio was maintained at 1.67 that is close to the molar ratio of the mineral phase in human bone [40]. Reagent grade chemicals of $Ca_2P_2O_7$, $Ca(OH)_2$, SiO_2, Na_2CO_3 and Fe_2O_3 in the appropriate proportions for each system were thoroughly mixed, sieved through a 125 μm sieve, and placed in a corundum crucible for the melting process in an electric furnace. The crucible was held at 800 °C for three hours of calcination. Then the temperature was increased at a rate of 5 °C/min to 1450 °C where it was held for the melting process for 30 min. The melt was quenched by pouring onto a stainless steel plate at RT. Subsequently, pieces from the four sample series were heat-treated in air for 6 hours at temperatures between 600 and 1100 °C. The specimens were named according to the starting composition in Fe_2O_3 and the heat treatment temperature, as listed in Table I.

Table I. Processing parameters and sample names in the four series according to their initial Fe_2O_3 content and heat-treatment temperatures.

Chemicals: $Ca_2P_2O_7$, $Ca(OH)_2$, SiO_2, Na_2CO_3, Fe_2O_3
Molar % of the oxides: $45(CaO.P_2O_5)$ $(52\text{-}x)SiO_2.xFe_2O_3.3Na_2O$
Ca/P molar ratio: 1.67

Heat treatment temperature T_A (°C)	Series 5G $x = 5$	Series 10G $x = 10$	Series 15G $x = 15$	Series 20G $x = 20$
As-prepared	5G	10G	15G	20G
600	5G600	10G600	15G600	20G600
800	5G800	10G800	15G800	20G800
900	5G900	10G900	15G900	20G900
1000	5G1000	10G1000	15G1000	20G1000
1100	5G1100	10G1100	15G1100	20G1100

XRD measurements were performed using a Siemens D5000 powder diffractometer operating at 45kV and 40mA with Ni-filtered Cu-Kα radiation and a diffracted beam monochromator. The data bank from the International Center for Diffraction Data (ICDD) was used in a search/match program for phase identification. The Rietveld refinement method in the GSAS program was used for crystal structure determination and quantitative analysis of phase fractions from the diffraction patterns.

Two major phases were identified in all sample series: calcium phosphate, $Ca_3(PO_4)_2$, crystallizing in the hexagonal and monoclinic systems, and magnetite (Fe_3O_4). Calcium silicate ($CaSiO_3$) that has been reported [7] as a major phase (wollastonite) in FBC systems of different composition was detected within instrumental detection limits with peaks overlapping with those of the major phases. Apparently, it did not crystallize under the processing conditions and remained in the amorphous matrix of the material. Also, Na_2O did not form a detectable crystalline phase. Fe_2O_3 appears as a secondary phase in the series 20G, while part of the SiO_2 of the starting chemicals is found as a secondary crystalline phase in some samples of series 10G.

The fractions of the crystalline phases in each specimen were determined by using standard Rietveld refinement techniques, as implemented by the GSAS program. Scale factors, background parameters, full width at half maximum profile parameters, and lattice parameters of the input phases were refined. The atomic parameters and temperature factors were not allowed to vary because of the complexity of the diffraction patterns and the input models. The data were analyzed solely for the crystalline phases and normalized to 100 wt %, while the amorphous content was ignored. The weight fraction W_p of a phase P in a specimen is given by [41]:

$$W_p = \frac{S_p(ZMV)_p}{\sum_i S_i(ZMV)_p}$$

Where S_p is the refinable scale factor for the phase P in the specimen; Z, M and V are the number of formula units per unit cell, the mass of the formula unit and the unit-cell volume of each phase respectively.

Figure III.1 shows an example of a Rietveld refinement of the diffraction pattern of the three-phase sample 15G800. Hexagonal calcium phosphate (SG R3c), monoclinic calcium phosphate (SG P21/a) and orthorhombic magnetite (SG Imma) were introduced in the calculated pattern. The amorphous content was not quantified.

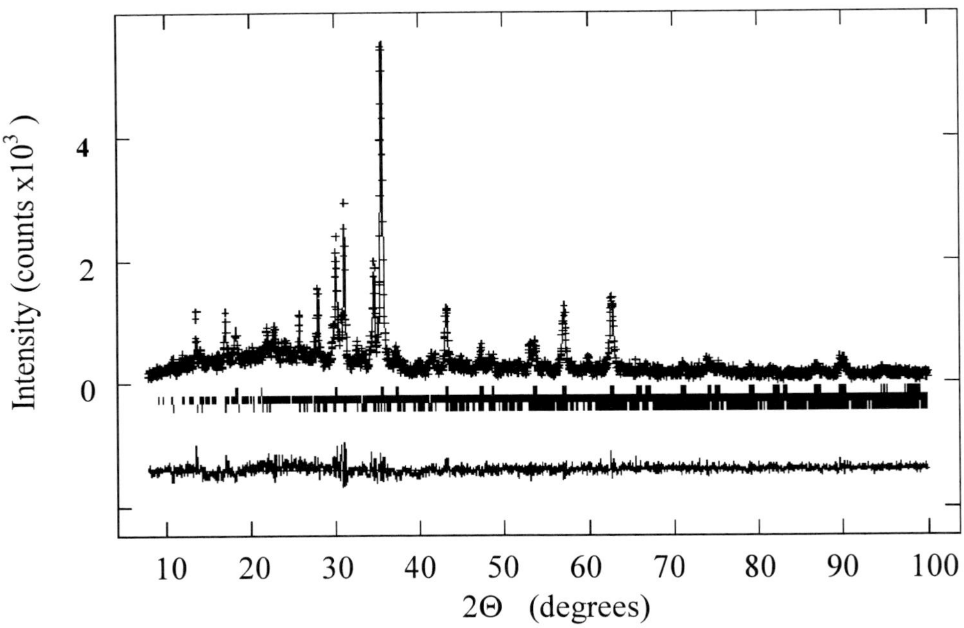

Fig. III. 1. Three-phase Rietveld refinement of the x-ray powder diffraction pattern of the sample 15G800. Crosses show the experimental pattern and solid line marks the calculated one. The first row of tick marks below the profiles shows the calculated Bragg peak positions for magnetite (SG Imma), while the second and third rows mark the ones for monoclinic (SG P21/a) and hexagonal (SG R3c) crystal structures correspondingly. The lowest trace is the difference between the experimental and calculated patterns.

Development of the crystalline phases in the samples series 10G, 15G and 20G is presented in Fig. III. 2 in which the refined phase fractions are plotted as a function of the heat-treatment temperature. Notice that in the sample series 10G calcium phosphate crystallizes only in the hexagonal system. The fraction of the minor phase of monoclinic SiO_2 (SG Cc) that remains from the starting chemicals in the untreated sample 10G increases in samples heat-treated at temperatures ≥1000 °C.

A gradual phase transition of calcium phosphate from P21/a→R3c is observed in series 15G and 20G for heat-treatment temperatures above 800 °C. A small fraction of monoclinic phase remains in samples annealed at 1000 °C or higher. Interestingly, the fraction of monoclinic calcium phosphate increases in samples heat-treated at 600 °C.

A fourth phase of Fe_2O_3 (SG R$\bar{3}$c) was introduced in the calculated patterns of the samples in series 20G, except for the one heat-treated at 600 °C. As Fig. III.2 shows, while the percentage of magnetite in series 15G is affected by the annealing temperature, the effect becomes dramatic in series 20G in which the magnetite weight fraction decreases by ~50 % from its maximum value in samples heat-treated at 1100 °C accompanied by formation of Fe_2O_3. The transformation of ferrimagnetic magnetite to non-magnetic Fe_2O_3 is detrimental to the properties of the material.

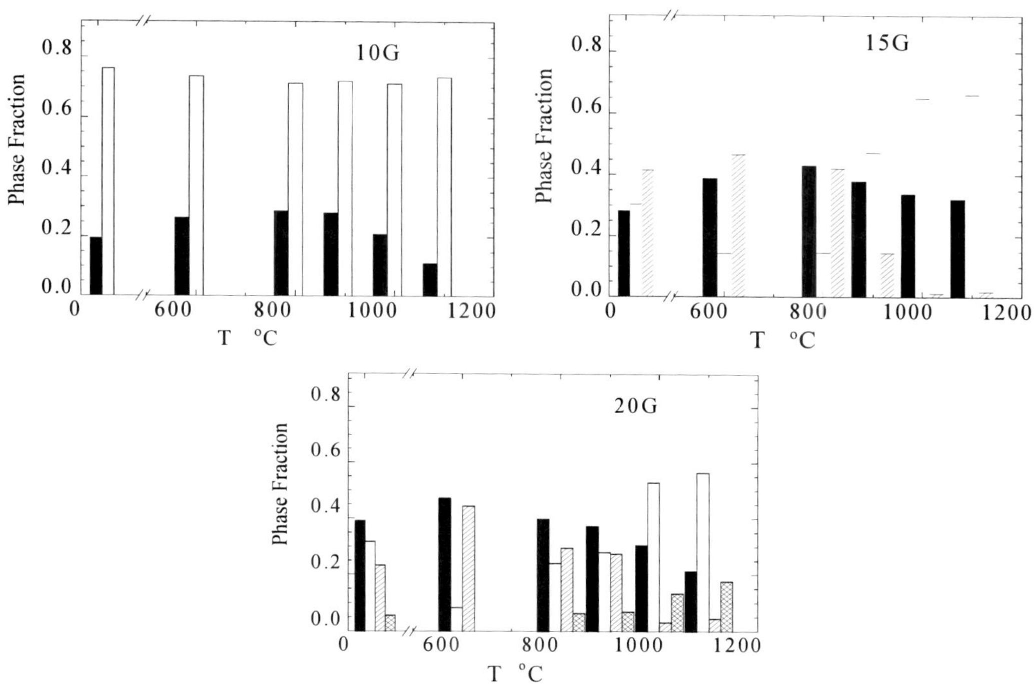

Figure III. 2. The refined phase fractions in the samples series 10G, 15G and 20G as a function of the heat-treatment temperature. Black marks the magnetite, white the hexagonal calcium phosphate, stripes the monoclinic calcium phosphate, and a diamond pattern the Fe_2O_3.

Scanning electron microscopy (SEM) with energy dispersive X-ray spectroscopy (EDX) was used for microstructural and microchemical analysis of the composite specimens. SEM back-scattered electron (BSE) imaging was performed with 15 keV incident electrons in order to acquire images with compositional contrast. EDX was performed on various phases for qualitative analysis of phase compositions and phase identification. Spectra were acquired at lower incident electron energies (e.g. 4 keV) as necessary to achieve improved spatial resolution. The SEM images have revealed a strong variation of microstructure with composition and heat treatment. The variation of the as-quenched microstructure with increasing iron oxide concentration is shown in Figure III. 3. Apart from the inset in Fig. III. 3c, the images are of identical magnification. The specimen with the lowest iron oxide concentration (5 wt% Fe_2O_3) is shown in Figure III.3a, which features a bimodal isotropic microstructure. EDX spectra indicate that the coarse precipitates primary shown in the upper right corner are pure silica, the darkly imaging secondary precipitates are calcium phosphates, and the brightly imaging glassy matrix contains all of the elements present in the starting mixtures. The primary silica precipitates are also present as darkly imaging features in Figure III.3b, but at this higher iron concentration (10 wt % Fe_2O_3), brightly imaging iron oxide dendrites also appear. For specimens with 15 wt % Fe_2O_3, the microstructure is featureless at the magnification used to image the specimens, but at higher magnification (inset), a fine-scale homogeneous phase distribution is revealed. The homogeneity of the microstructure is maintained at the highest iron concentration (20 wt% Fe_2O_3), but a more coarse dendritic structure is achieved.

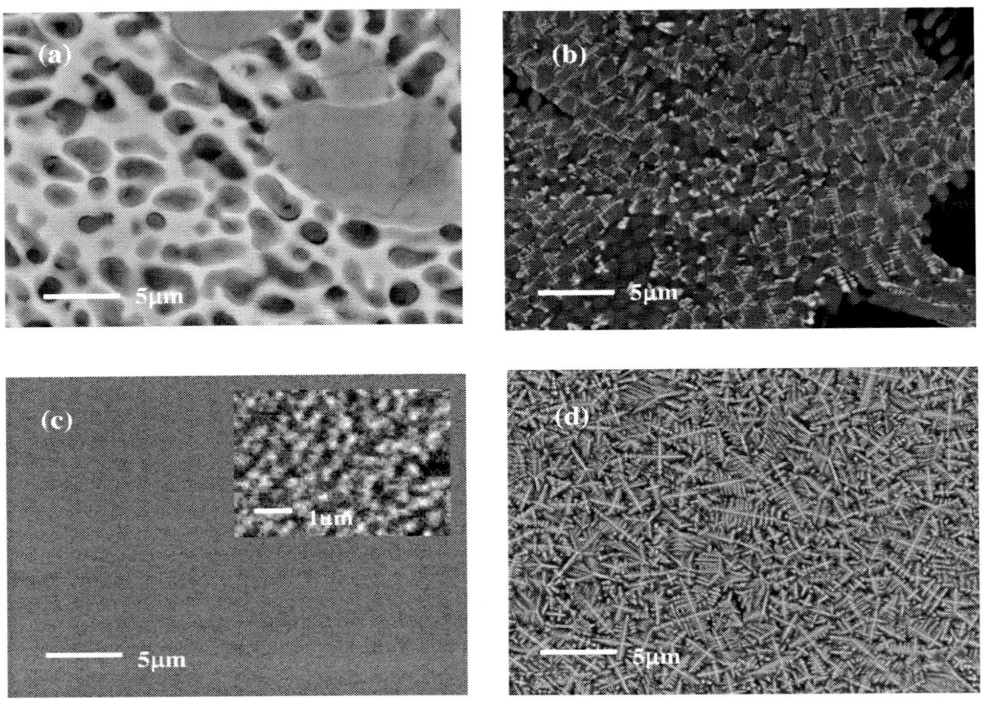

Figure III. 3. SEM images illustrating the composition dependent evolution of the microstructures in the as-prepared FBC materials: **(a)** sample 5G, **(b)** sample 10G, **(c)** sample 15G, and **(d)** sample 20G.

Evolution of the dendritic microstructure (shown in Figure III.3d) with heat-treatment temperature is illustrated in Figure III.4. The magnification of the images is twice that in Figure III.3. The dendrite grains of the unannealed sample Figure III.4a become longer with oriented domains at 600 °C (Figure III.4b), and 800 °C (Figure III.4c). However, in samples heat-treated at 900 °C or higher, the long dendrite structures start to "break up" and, at 1100 °C, no well-defined dendritic structure is observed (Figure III.4d), although one can still see a dendrite trace with a distinct phase separation.

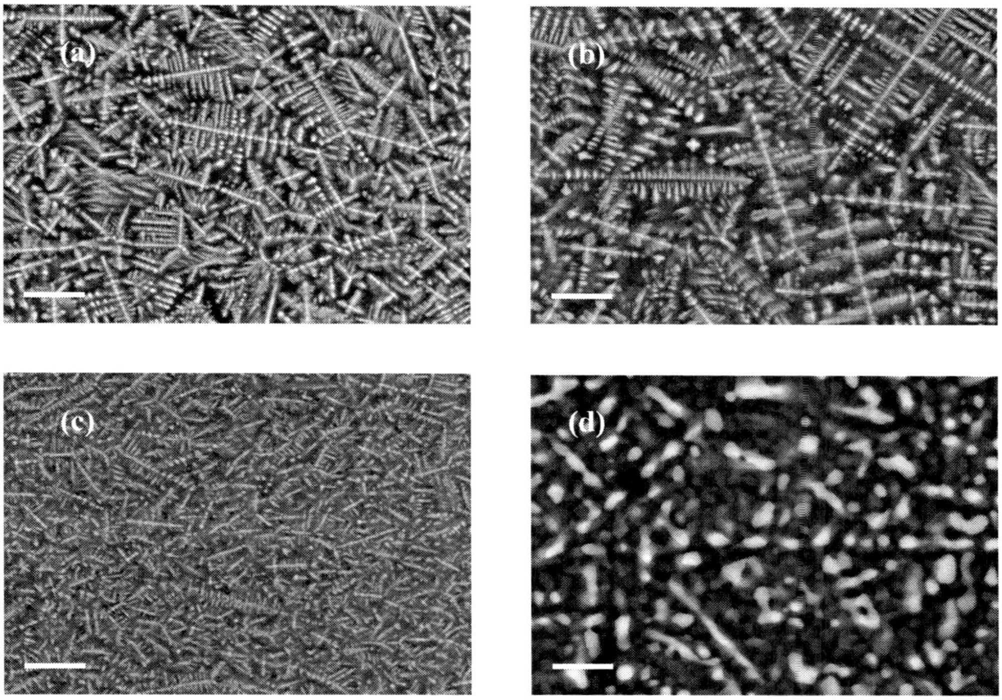

Figure III. 4. SEM images from samples of the series 20G presenting the effect of heat-treatment temperature on the microstructure of FBC: (a) as-prepared, (b) heat-treated at 600 oC, (c) at 800 oC and (d) at 1100 oC. Bar is 2 □m.

Let us now consider the magnetic properties of the materials. Fig. III. 5 illustrates the compositional dependence of the magnetization of samples that were prepared under similar processing conditions with varying initial molar concentration x of Fe_2O_3, with x=0.05, 0.10, 0.15, and 0.20. Room temperature *M-H* curves with applied field up to 4000 Gauss for the as-prepared materials are plotted in this figure. Notice that the magnetization of the samples 15G and 20G appears similar, while the crystalline phase fraction of magnetite in sample 20G is ~28% greater than in the 15G (Fig. III. 2); this is probably associated with the very fine structure observed in sample series 15G (Fig. III. 3c) compared with the large dendrites in series 20G (Fig. III. 3d).

An example of the effect of heat-treatment at elevated temperatures on the magnetic properties of the biomaterial is illustrated in Fig. III. 6, where magnetization loops for four out of the six samples in series 20G are plotted; two loops are omitted for reasons of clarity. The loop area initially increases with annealing temperature T_A when the untreated sample 20G (marked with open circles) is heat-treated at 600 °C, (20G600, filled circles). The loop area and magnetization decrease to a lower value for T_A = 1000 °C (squares); this reflects the conversion of ferrimagnetic Fe_3O_4 into nonmagnetic hematite that is evident in Fig. III. 2. These magnetic quantities decrease still more in samples heat-treated at 1100 °C (open squares) due to additional conversion of magnetite into hematite.

Large values of remanence M_r and coercivity H_c are desirable for the FBC, since the hysteretic ferrimagnetic loss determines the hyperthermic properties of the biomaterial. A correlation of the parameters *Ms* with the molar composition of the reacting oxides and heat-treatment temperature is illustrated in Figure III.7 for all the samples studied.

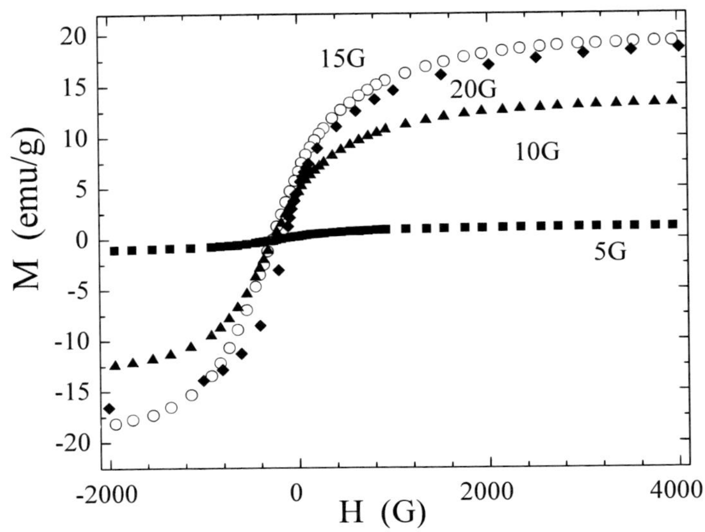

Figure III. 5. M-H plots of the as-prepared samples with 5, 10, 15 and 20% molar fraction of Fe2O3 in the reacting oxides. The response in applied magnetic field H up to 4 kG is shown.

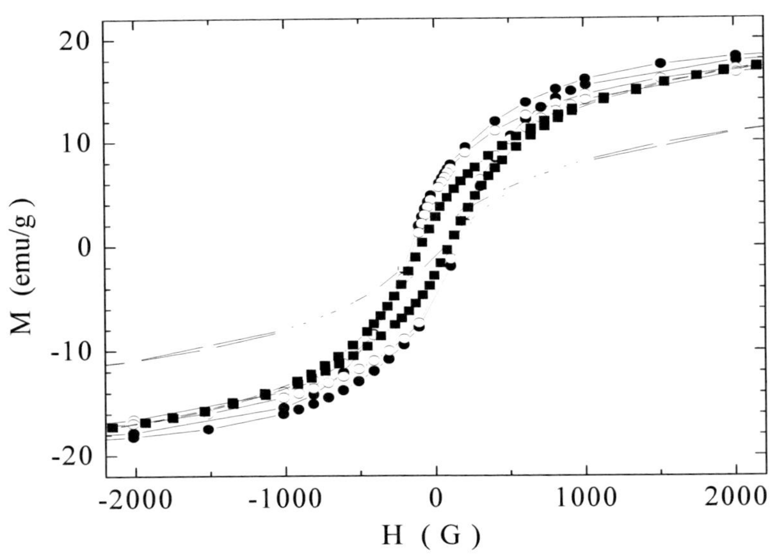

Figure III. 6. Hysteresis loops of the samples in series 20G heat-treated at temperatures from 600-1100 °C. Open circles mark the as-prepared sample 20G, filled circles the 20G600, squares the 20G1000, and open squares the 20G1100. *M-H* measurements were conducted at room temperature in magnetic fields up to 20 kG.

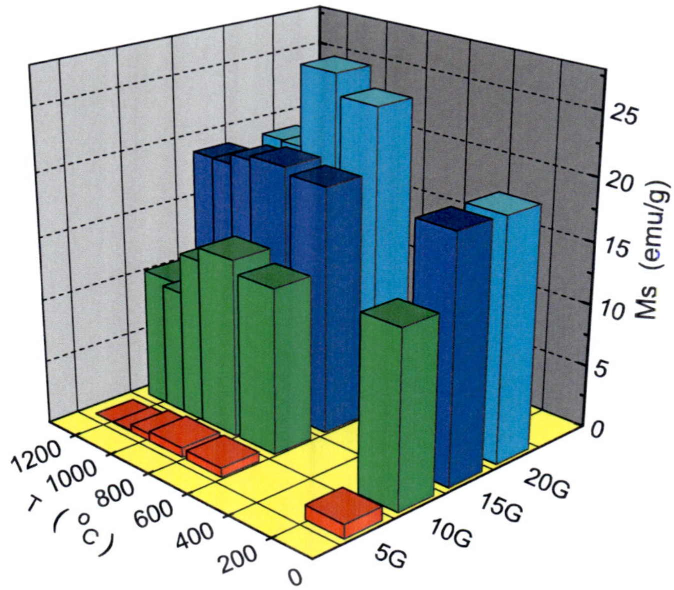

Figure III. 7. The saturation magnetization Ms and the refined Fe3O4 fraction in the four samples series as a function of heat-treatment temperature. Series 5G is marked with red columns, 10G with green, 15G with blue, and series 20G with light blue.

Notice that the saturation magnetization M_s in the samples of series 15G (blue columns) remains almost unaffected by heat-treatment, it is reduced in series 10G (green columns) by annealing above 900 °C, while it is reduced to ~55% of its maximum value in the samples of series 20G that were heat-treated at 1100 °C (light blue).

As the plot of the coercivity H_c versus heat-treatment temperature in Figure III.8 shows, materials with the highest magnetite fraction and M_s values (system 20G) become demagnetized (M = 0) at the lowest values of reverse applied magnetic field. For the 20G series, only a small effect of temperature T_A on H_c is observed. On the other hand, the system 15G attains higher values of H_c, which increases with annealing temperature. A similar, but even more pronounced trend for the H_c is revealed in the system 10G, which attains the largest coercivity. In general, the relatively large values of magnetite coercivity, 100-350 Oe compared with bulk materials, can be associated with the small particle size [42] of fine and/or "broken" dendrites of Fe_3O_4. Magnetite particles that are roughly equiaxed are expected to be single magnetic domains for sizes less than ~ 0.3 – 2 μm [43].For Fe_3O_4 particles in this size range, which were once considered for magnetic recording purposes, the values of H_c are comparable with those observed here [24]. Furthermore, an elongated particle geometry typically enhances magnetic hysteresis, and this may explain why H_c is larger in the series 10G and 15G than in the 20G materials, in which the "breaking" of dendrites reduces the shape anisotropy of the magnetite particles.

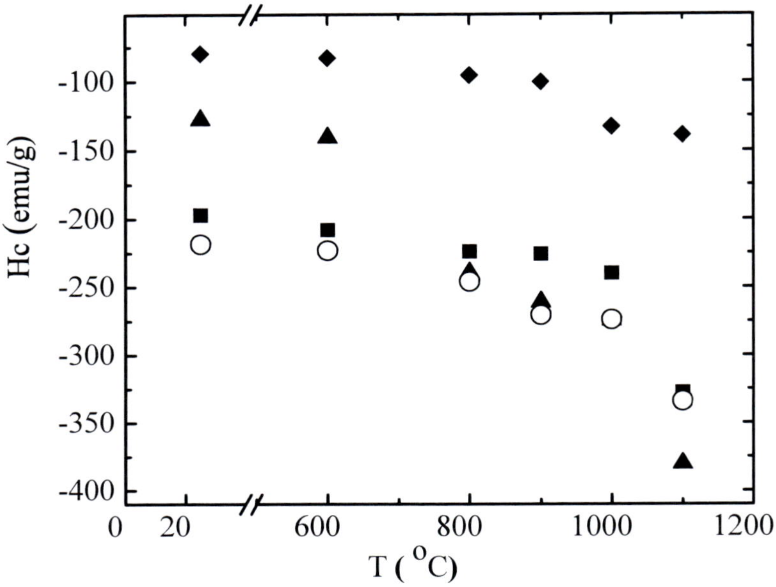

Figure III. 8. The coercivity Hc of the four samples series as a function of heat-treatment temperature. Series 5G is marked with filled squares, series 10G with triangles, series 15G with open circles, and series 20G with diamonds. Error bars in plots are of the size of the markers

The magnetic remanence M_r that reaches its highest values for the samples in series 15G at all temperatures as Figure III.9 shows, can also be associated with the fine dendrites of magnetite in these samples (Figure III.3c). Annealing the series 10G and 20G at T_A = 1000 – 1100 °C diminishes the remanence significantly.

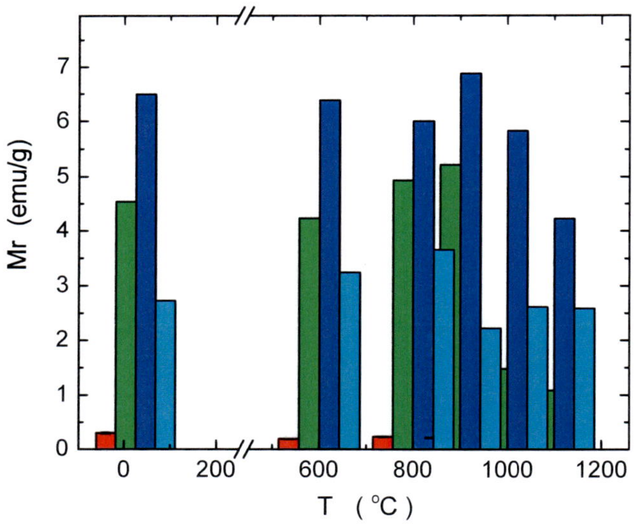

Figure III. 9. The remanence *Mr* of the four samples series as a function of heat-treatment temperature. Series 5G is marked with red columns, series 10G with green, series 15G with blue, and series 20G with light blue.

At this point, it is useful to relate the observed saturation magnetization M_s with the compositional analysis from the Rietveld analysis. Their relation is shown in Fig. III. 10, which displays M_s (in emu per gram of sample) as a function of the Fe_3O_4 fraction of the total crystalline mass. For comparison the solid line in the figure shows the magnetization expected for bulk crystalline magnetite, ~90 G cm^3/g [42]. As expected, the experimental magnetization generally increases with the magnetite fraction increases. However, the present data lie below the level nominally expected from bulk Fe_3O_4, which in fact appears to be an upper bound on the observed values. This difference can be attributed in large part to the presence of glassy phases in the materials that are not included in the Rietveld analysis, but which constitute part of the overall sample mass used to calculate the saturation magnetization. This interpretation is consistent with a significant presence of non-crystalline material in most of the samples, which is not surprising for bioglass ceramics.

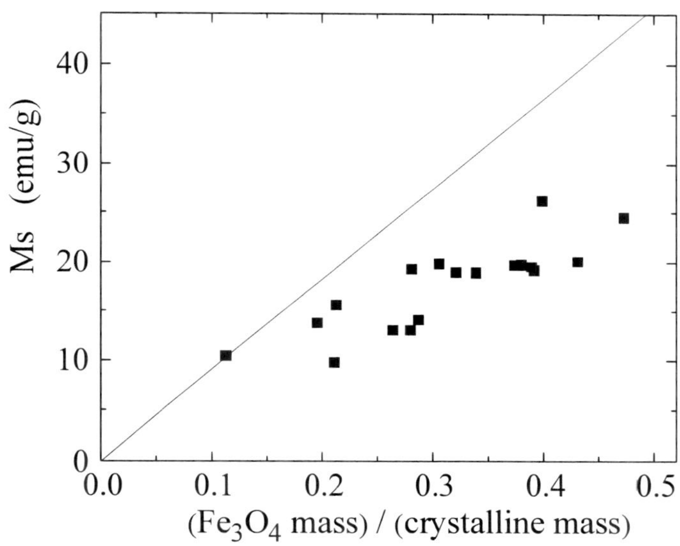

Figure III.10. The saturation magnetization plotted versus the magnetite content in all the sample series. The solid line shows the "ideal" result for bulk Fe3O4 in a completely crystalline multiphase material.

The results of the present studies show that the effect of synthesis variables on the structure and magnetism of FBC are bound to the specific parameters of the sample series. For instance, while the saturation magnetization M_s is clearly determined by the concentration of the reacting oxides in Fe_2O_3 up to 15%, heat-treatment temperature takes over in sample series 20G as demonstrated in Fig. III.7. As the SEM image of Fig. III.4d illustrates, the breaking of the ferrimagnetic dendrites above 800 °C correlates with the large changes in the magnetic properties of the material. These changes also correlate with a progressive conversion of magnetite into Fe_2O_3, whose (crystalline) weight fraction increases from 0 in the sample 20G600 to 0.17 in the 20G1100 (Fig. III. 2). Notice in the same figure that only the sample heat-treated at 600 °C contains no Fe_2O_3 while Fig. III.6 reveals that the maximum magnetic hysteresis loop develops in this specimen. The maximum value of the saturation magnetization in these samples reaches 26 emu/g, almost half from the expected value (≅ 40 emu/g) for a sample containing 45% magnetite. The reported M_S values for pure magnetite are about 90-92 emu/g [44,45]. Such lower M_S indicates that our specimens contain ferrimagnetic phase with some magnetic "impurities" within a glassy matrix that was not included in the Rietveld analysis of the XRD patterns. These results are in agreement with the results of the quantitative Rietveld refinement analysis of the x-ray diffraction patterns that have shown a certain percentage of hematite in addition to the magnetite (Fig. III.2).

From the application point of view, the energy stored in the magnetization loops of the FBC is the most interesting property of the materials. The heating power of specimens of the four starting compositions, heat-treated at 1000 °C, and of all specimens in series 20G was calculated from the corresponding magnetization loops and for alternating magnetic fields of 100 kHz as a measure of their heating ability; the results are plotted in Figure III.11. Circles are used for the sample series 20G that were heat-treated from 600 up to 1100 °C and triangles for each sample from the other series that was heat-treated at 1000 °C. The values reach ~200 W/g, which is comparable to the ones reported for systems of different composition [46].

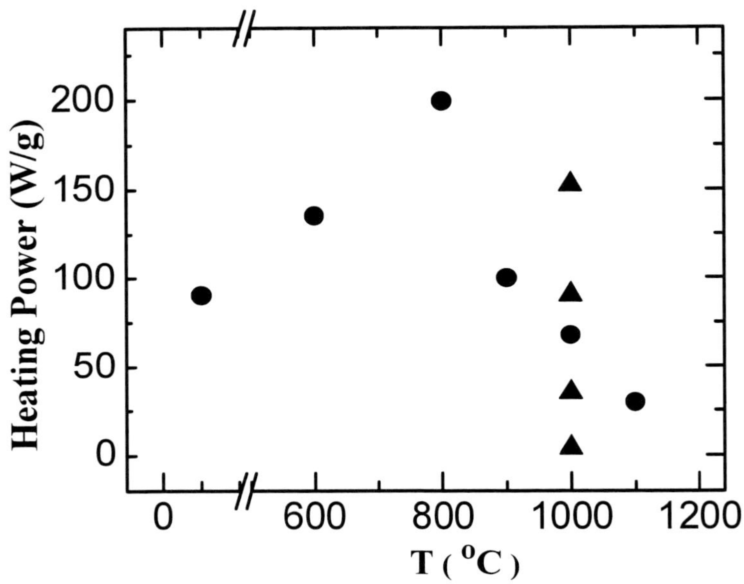

Fig. III. 11. Heating power per gram as a function of the heat-treatment temperature and starting composition. It is calculated from the corresponding magnetization loops and for a 100 kHz frequency of applied magnetic field. Samples of series 20G are marked with circles. Triangles correspond to the samples 5G1000, 10G1000, and 15G1000 with increasing values of heating power from 5G to 15G.

In vivo experiments are necessary to quantify the heating power and evaluate a potential application of these materials in local hyperthermic treatment of bone cancer. As a first step we have started with *in vitro* experiments to demonstrate formation of hydroxyapatite and pores when our samples are immersed in a simulated body fluid. Experimental studies of the crystal structure and surface structure are discussed in part IV of this Chapter.

IV. *In Vitro* Studies of Ferrimagnetic Bioglass Ceramics

The effect of simulated body fluid (SBF) on the structure and microstructure of ferrimagnetic bioglass ceramics (FBC) was investigated in series of samples in the system of the oxides [0.45(CaO, P_2O_5) (0.52-x)SiO_2 xFe_2O_3 0.03Na_2O], with x = 0.05, 0.10, 0.15, 0.20. Simulated body fluid (SBF) was prepared by dissolving reagent grade chemicals NaCl, $NaHCO_3$, KCl, $K_2HPO_4.3H_2O$, $MgCl_2.6H_2O$, $CaCl_2$, Na_2SO_4; the solution was buffered at pH 7.25 with 50 mM tris [$(CH_2OH)_3CNH_2$] and 45 mM of hydrochloric acid at 37 °C (human body temperature) [39]. The ion concentrations in SBF are given in Table I along with the ones for human blood plasma.

Table I. Ion concentrations (mM) of simulated body fluid (SBF) and human blood plasma (HBP).

	Na^+	K^+	Mg^{2+}	Ca^{2+}	Cl^-	HCO_4^{2-}	HPO_4^{2-}	SO_4^{2-}
SBF	142.0	5.0	1.5	2.5	147.8	4.2	1.0	0.5
HBP	142.0	5.0	1.5	2.5	103.0	27.0	1.0	0.5

Table II lists the FBC samples in these experiments. Crystallographic and surface structures of the materials were studied as a function of composition, heat-treatment temperature, and time of immersion in SBF by using x-ray diffraction (XRD) and scanning electron microscopy (SEM) with energy dispersive x-ray spectroscopy (EDS). For the reader's convenience we give an example of what the name of each specimen means: sample 15G100040 refers to a sample from the series 15G with 15% Fe_2O_3 in the starting composition of the oxides, heat treated at 1000 °C, and immersed in SBF for 40 days.

Table II. Ferrimagnetic Bio Ceramic (FBC) samples of various compositions in starting oxides, heat treatment temperature, and days of immersion in SBF.

Chemicals: $Ca_2P_2O_7$, $Ca(OH)_2$, SiO_2, Na_2CO_3, Fe_2O_3 Molar % of the oxides: 45($CaO.P_2O_5$) (52-x)$SiO_2.xFe_2O_3.3Na_2O$				
Time in SBF (days)	Series 5G $x = 5$	Series 10G $x = 10$	Series 15G $x = 15$	Series 20G $x = 20$
0	5G	10G	15G	20G
8	5G8	10G8	15G8	20G8
18	5G18	10G18	15G18	20G18
25	5G25	10G25	15G25	20G25
32	5G32	10G32	15G32	20G32
"	-	-	-	20G90032
"	-	-	-	20G100032
40	5G40	10G40	15G100040	20G40
115	-	-	15G115	-
123	-	10G123	-	20G123

Figure IV.1 shows the evolution of crystalline phases versus time in SBF and heat treatment temperature. On the left, part of the diffraction pattern of the as-prepared sample 15G is shown in comparison with the pattern of 15G18 (18 days in SBF). In the middle: 15G is compared with the 15G115 (115 days in SBF). On the right part of the Figure the diffraction pattern of the 15G is compared with the one from 15G100040; it also shows the evolution of the crystalline phases in a sample that was annealed at 1000 °C prior to immersion in SBF for 40 days. For clarity reasons, only the Miller indices of the major hydroxyapatite (HAp) peaks are marked on the diffraction patterns.

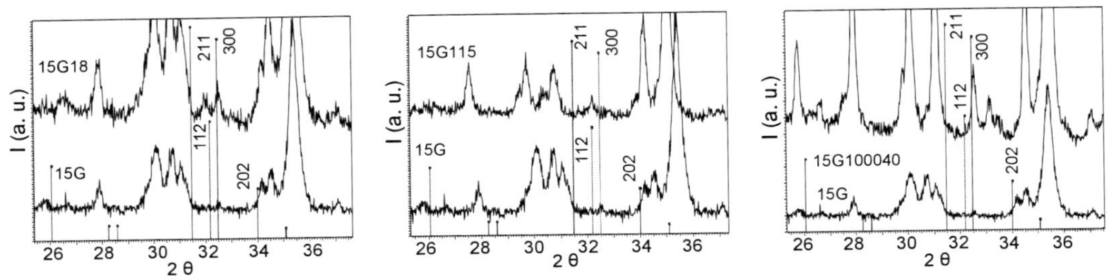

Figure IV.1. Part of the XRD pattern of the as-prepared sample 15G in comparison with the one from the specimen 15G18 (18 days of immersion in SBF, on the left), and with the one from the 15G115 (115 days in SBF, middle). The diffraction pattern of 15G is also compared with the one from the sample 15G100040 that was annealed at 1000 °C, and then immersed in SBF for 40 days (right). The hydroxyapatite peak positions and corresponding *h k l* indices are marked on the patterns.

The bioactive hydroxyapatite phase (ICDD # 70-0795) develops in 15G18, 15G115, and 15G100040. Notice that the relative intensities of the HAp peaks are not any higher in the diffraction pattern of the specimen 15G115 that was immersed for 115 days in SBF, infact, these are lower than the ones in 15G18, indicating that prolonged immersion time does not increase the bioactivity of the material. Similar conclusions have been reported by others [47].

In addition to bioactivity, exposure of the samples to the ions of SBF, and time of immersion affect the structural evolution of all major phases in the materials: magnetite is identified in the cubic crystal system in all three specimens of Figure IV.1 after exposure in SBF; calcium phosphate crystallizes in the hexagonal and orthorhombic systems in the specimen 15G18, in the orthorhombic system in the specimen 15G115, while in the specimen 15G100040 it appears only in the hexagonal system. Calcium silicate (cyclowollastonite) in the triclinic system was identified only in the specimen 15G115.

The relative intensities of the HAp peaks of the diffraction patterns on the right part of the Figure indicate that annealing of a specimen from the 15G series at 1000 °C, followed by immersion in SBF for 40 days, results to an increased bioactivity of the material.

Scanning electron microscopy (SEM) with energy dispersive x-ray spectroscopy (EDX) was used for microstructural and microchemical analysis of the composite specimens. SEM back-scattered electron (BSE) imaging was performed with 15 keV incident electrons in order to acquire images with compositional contrast. EDX was performed on various phases for qualitative analysis of phase compositions and phase identification. Spectra were acquired at lower incident electron energies (e.g. 4 keV) as necessary to achieve improved spatial resolution.

Figures IV.2-5 display the effect of SBF on the microstructure of these materials. Notice that the iron, oxygen dendrites are largely covered by a layer consisting of silicon, calcium, phosphorus, sodium and oxygen; also notice formation of the desirable pores that are considered a possible hydroxyapatite nucleation site therefore increase the bioactivity of the materials [47,48].

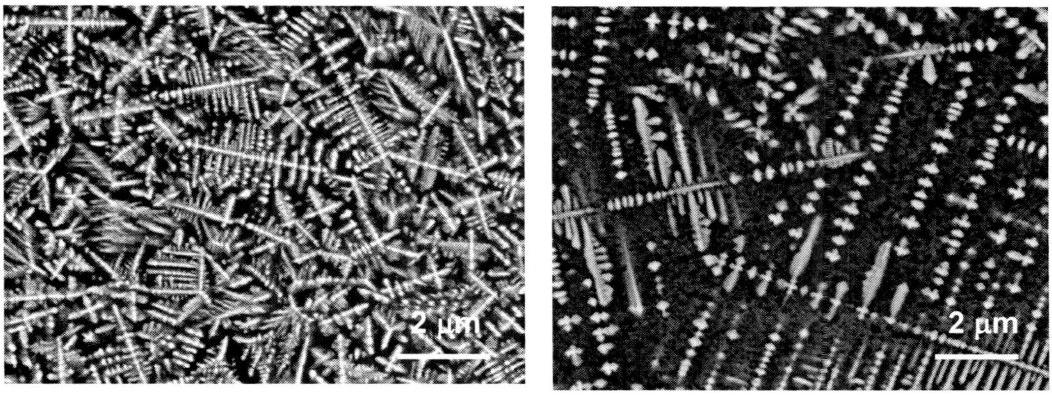

Figure IV.2. BSE SEM image of the as prepared sample 20G on the left, in comparison with the image of the same specimen after it was immersed in SBF for 40 days (20G40) on the right. The dendrites of the 20G are now largely covered by a layer, while pores of various sizes form on the surface of 20G40.

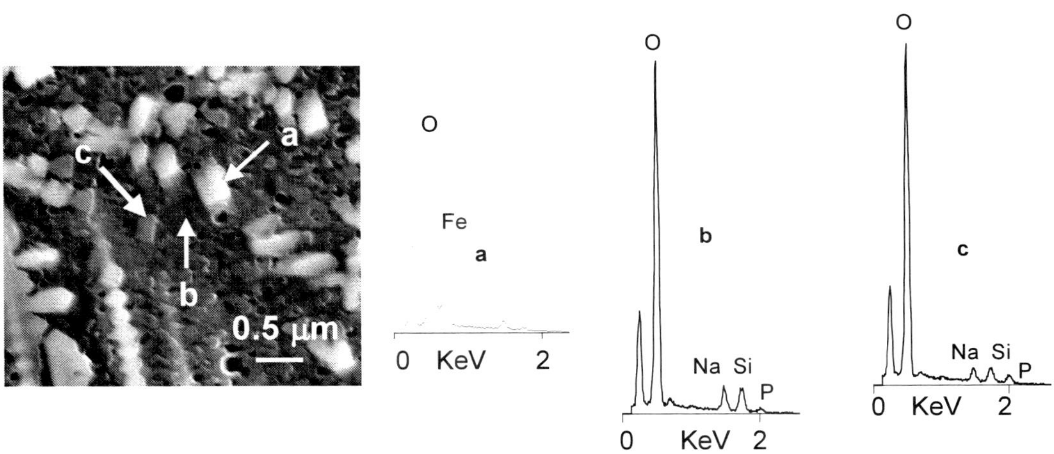

Figure IV.3. EDX spectra from the sample 20G40: **(a)** the bright crystalline structures that form long dendrites mainly consist of oxygen and iron, **(b)** the dark layer next to them contains high percentage of oxygen, silicon and sodium, **(c)** oxygen, silicon, sodium, phosphorus and calcium (not shown in the spectrum) are distributed on the layer and the glassy phase between the crystallites. Notice the sizes of the pores on the specimen. The peak to the left of oxygen is carbon, which was used to coat the specimens.

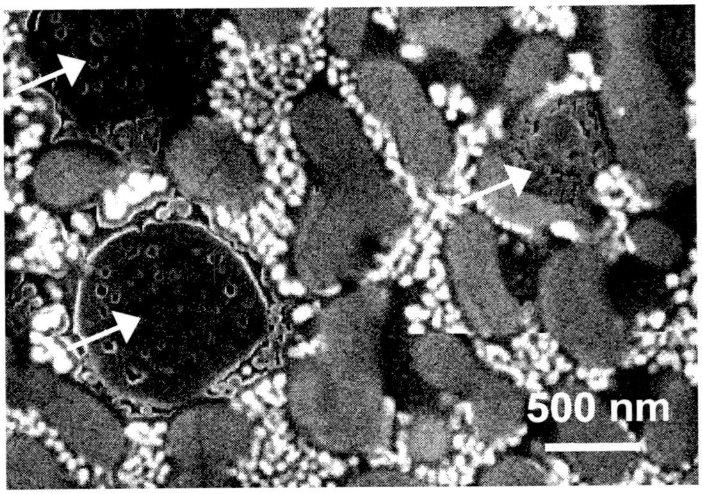

Figure IV.4. SEM image of the specimen 10G40. Nanoscale pores develop on the dark and gray phases.

A comparison between Figures IV.4 and IV.5 demonstrates that while porous structures develop in the sample series with starting molar composition 10% in Fe_2O_3, the size of the pores is about three orders of magnitude smaller in the 10G40 of Figure IV.4 than in the series 20G with 20% Fe_2O_3, Figure IV.5 on the left, specimen 20G90032. The effect of the heat-treatment temperature on pore formation is demonstrated in Figure IV.5. The one on the right, demonstrates that annealing at temperatures above 900 °C, does not allow formation of pores on the surface of the biomaterial. In the case of the series 15G the surface structure was so fine, that neither phases were resolved nor pores were observed.

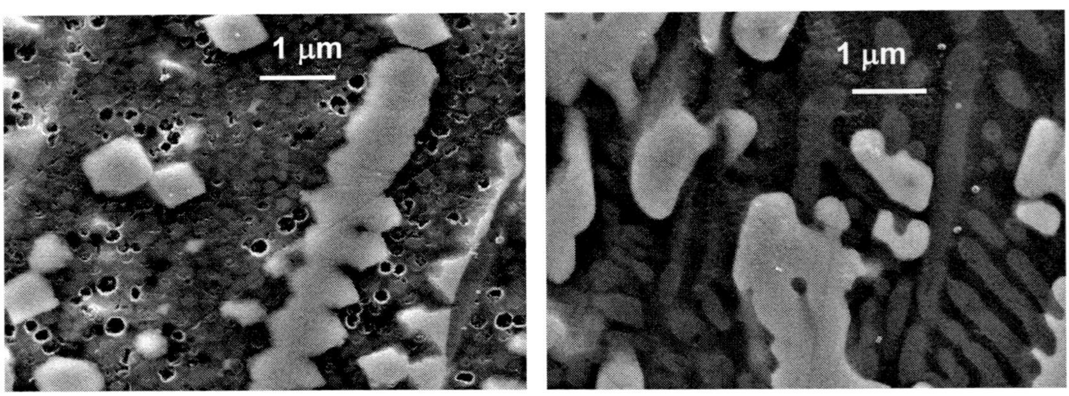

Figure IV.5. SEM images of the specimens 20G90032 (left) and the 20G100032 (right). Both specimens were immersed in SBF for 32 days, but the one where pores develop was annealed at 900 °C, while the other that does not form any pores was annealed at 1000 °C.

Overall, the results of the *in vitro* experiment have demonstrated a non-systematic variation of the bioactivity of FBC with the composition in starting oxides, heat treatment temperature, and time of exposure in SBF. A surface layer of Si, P, Ca, O and Na partially cover the Fe, O dendrites, and pores form in various sizes determined by the specific processing parameters of the materials.

It is obvious that more controlled experiments (e.g. tracking of a single area of the microstructure with increasing exposure to SBF) might be required to establish a definite correlation of microstructural evolution with SBF exposure. Further studies on associations between processing parameters and bioactivity, as well as evaluation of the magnetic properties of the biomaterials after soaking in SBF, could provide additional insights regarding the application of FBC for hyperthermic treatment of deep tissue cancer. Use of high-resolution neutron powder diffraction (NPD) to study qualitatively and quantitatively the development of crystallographic phases versus processing parameters in magnetic biomaterials can provide detailed information, since light elements have large neutron scattering lengths as oppose to small x-ray scattering factors.

Another venue is to research different magnetic materials: Magnetic approaches are particularly nice because they involve biocompatible materials as well as non-invasive heating. The limitations are that iron oxides, such as magnetite, are not optimum materials for delivering energy through hysteretic loss and magnetic properties of these oxides strongly depend on processing conditions. Using magnetic materials with greater coercivity, higher magnetization, and more magnetically anisotropic crystal structure could drastically enhance the efficiency of energy delivery by "spreading" the hysteresis loop. In addition, different magnetic materials could lead to more predictable behavior under a variety of processing conditions.

Conclusion

The systematic studies of crystal structure and microstructure of hydroxyapatite based biomaterials discussed in this chapter clearly demonstrates that a considerable amount of knowledge has been achieved. At the same time, it has been brought out the fact that many questions remain unresolved. More basic research is required in both, macro and nanoscale biomaterials in order to achieve quality products with controlled properties.

Acknowledgements

I am gratefully thankful to my colleagues Drs. B.C. Chakoumakos, V. Perdikatsis, J. R. Thompson, I. M. Anderson, and students Dr. A. C. Kis, Dr. N. Papanearchou, C. Bunaciu, and H.Y. Moghaddam who have made the research project on HAp-based biomaterials possible over the last few years. Research supported by the Southeastern Universities Research Association (SURA\ORNL) Oak Ridge National Laboratory, and the Cancer Institute at the FAU Research Park, Boca Raton, FL.

REFERENCES

[1] Gimcher, MJ; Bonar, LC; Grynpas, MC; Landis, WJ; Roufosse, AH. Recent studies of bone mineral: Is the amorphous calcium phosphate theory valid? *J. Cryst. Growth*, (1981) 53, 100-119.

[2] Zapanta LeGeros, R. Calcium Phosphates in Oral Biology and Medicine. New York: Karger; 1991.

[3] Elliott, JC. Structure and Chemistry of the Apatites and Other Calcium Orthophosphates. Amsterdam: Elsevier; 1994.

[4] Smith, DK. Title:Calcium phosphate apatites in nature. In: Brown PW, Constantz B, editors. *Title: Hydroxyapatite and related materials*. Boca Raton: CRC; 1994; 29-44.

[5] Carlisle, EM. Silicon: A possible factor in bone calcification. *Science*, (1969) 167, 279-280.

[6] Ikenaga, M; Ohura, K; Nakamura, T; Kotoura, Y; Yamamuro, T; Oka, M; Ebisawa, Y; Kokubo, T. Hyperthermic Treatment of Experimental Bone Tumours with a Bioactive Ferromagnetic Glass-Ceramic. *Bioceramics*, (1991) 4, 255-262.

[7] Ohura, K; Ikenaga, M; Nakamura, T; Yamamuro, T; Ebisawa, Y; Kokubo, T; Kotoura, Y; Oka, M. A Heat-Generating Bioactive Glass-Ceramic for Hyperthermia. *J. Appl. Biomat.*, (1991) 2, 153-159.

[8] Kokubo, T; Ebisawa, Y; Sugimoto, Y; Kiyama, M; Ohura, K; Yamamuro, T; Hiraoka, M; Abe, M. Preparation of bioactive and ferromagnetic glass-ceramic for hyperthermia. *Bioceramics*, (1992) 5, 213-223.

[9] Hench, LL; Splinter, RJ; Allen, WC; Greenlee, TK Jr. Bonding mechanisms at the interface of ceramic prosthetic materials. *J. Biomed. Mater. Res.*, (1972) 2, 117-141.

[10] Hench, LL; Wilson, JW. Surface-Active Biomaterials. *Science,* (1984) 226, 630-636.

[11] Grooss, U; Strunz V. The Interface of Various Glasses and Glass-Ceramics with a Body Implantation Bed. *J. Biomed. Mater. Res.*, (1985) 19, 251-271.

[12] Hench, LL; West JK. The Sol-Gel Process. *Chem. Rev.*, [1990] 90, 33-72.

[13] Cho, C; Wu, Y; Ackerman, JL. Detection of Hydroxyl Ions in Bone Mineral by Solid-State NMR Spectroscopy. *Science* (2003) 300, 1123-1126.

[14] Ohura, K; Ikenaga, M; Nakamura, T; Yamamuro, T; Ebisawa, Y; Kokubo, T; Kotoura, Y; Oka, M. A bioactive glass-ceramic for treatment of bone tumors. *Bioceramics* (1992) 3, 225-229.

[15] Ikenaga, M; Ohura, K; Yamamuro, T; Kotoura, Y; Oka, M; Kokubo, T. Localized Hyperthermic Treatment of Experimental Bone Tumors with Ferromagnetic Ceramics. *J Orthopaedic Res* (1993) 11, 849-855.

[16] Wilson, RM;. Elliott, JC; Dowker, SEP; Smith, RI. Rietveld structure refinement of precipitated carbonate apatite using neutron diffraction data. *Biomaterials,* (2004) 25, 2205-2213.

[17] Ivanova, TI; Frank-Kamenetskaya, OV; Kol'tsov, AB; Ugolkov, VL. Crystal Structure of Calcium- Deficient Carbonated Hydroxyapatite: Thermal Decomposition. *J. Solid State Chem.*, (2001) 160, 340-349.

[18] Fleet, ME; Liu, X. Location of type B carbonate ion in type A-B carbonate apatite synthesized at high pressure. *J. Solid State Chem.*, (2004) 177, 3174-3182.

[19] Leventouri, T; Chakoumakos, BC; Papanearchou, N; Perdikatsis, V. Comparison of crystal structure parameters of natural and synthetic apatites from neutron powder diffraction. *J. Mater. Res.*, (2001) 16, 2600-2606.
[20] Leventouri, T; Chakoumakos, BC; Papanearchou, N; Perdikatsis, V. Crystal Structure Studies of Natural and Synthetic Apatites from Neutron Powder Diffraction. *Mater Sci Forum*, (2001) 378, 517-522.
[21] Leventouri, T; Chakoumakos, BC; Moghaddam, HY; Perdikatsis, V. Powder neutron diffraction studies of a carbonate fluorapatite. *J. Mater. Res.*, (2000) 15, 511-517.
[22] Rietveld, HM. A Profile Refinement Method for Nuclear and Magnetic Structures. *J. Appl. Crystallogr.* (1969) 2, 65-71.
[23] Perdikatsis, V. X-ray powder diffraction study of francolite by the Rietveld method. *Mater. Sci. Forum* (1991) 79-82, 809-814.
[24] Larson, AC; Von Dreele RB. General Structure Analysis System. Los Alamos Nat. Lab. Rept. 1986, LAUR 86-748, version MS-DOS 3/2000.
[25] Sänger, AT; Kuhs, WF. Structural disorder in Hydroxyapatite. *Z. Kristallogr.*, (1992) 199, 123-148.
[26] Zapanta-LeGeros, R. Effect of Carbonate on the Lattice Parameters of Apatite. *Nature* (1965) 206, 403-404.
[27] Skinner, HCW. In Origin, Evolution and Modern Aspects of Biomineralization in Plants and Animals. Editor: Rex E. Crick Plenum Press; 1989.
[28] Baur, WH. The geometry of polyhedral distortions. Predictive relationships for the phosphate group. *Acta Cryst. B*, (1974) 30, 1195-1215.
[29] Ruys AJ. Silicon-doped hydroxyapatite. *J. Aust. Ceram. Soc.* (1993) 29, 71-79.
[30] Gibson, IR; Best, SM; Bonfield, W. Chemical characterization of silicon-substituted hydroxyapatite. *J. Biomed. Mater. Res.* (1999) 44, 422-428.
[31] Kim, SR; Riu, DH; Lee, YJ; Kim, YH. Synthesis and characterization of silicon substituted hydroxyapatite. *Key Engineering Materials* (2002) 218-220, 85-88.
[32] Leventouri, T; Bunaciu, CE; Perdikatsis, V. Neutron powder diffraction studies of silicon-substituted Hydroxyapatite. *Biomaterials* (2003) 24, 4205–4211.
[33] Baur, WH. The geometry of polyhedral distortions. Predictive relationships for the phosphate group. *Acta Cryst* (1974); B30: 1195–215.
[34] Takegami, K; Sano, T; Wakabayashi, H; Sodona, J; Yamazaki, T; Morita, S; Shibuya, T; Uchida, A. New Ferromagnetic Bone Cement for Local Hyperthermia. *J. Biomed. Mater. Res.* (1998) 43, 210-214.
[35] Singh, K; Bahadur, D. Characterization of SiO_2-Na_2O-Fe_2O_3-CaO-P_2O_3-B_2O_3 glass ceramis. *J. Mater. Sci: Mater. Med.* (1999) 10, 481-484.
[36] Salinas, AJ; Roman, J; Vallet-Regi, M; Oliveira, JM; Correia, RN; Fernandes, MH. In vitro bioactivity of glass and glass-ceramics of the 3CaO.P_2O_5CaO.SiO_2CaO.MgO.2SiO_2 system. *J. Biomat.* (2000) 21, 251-257.
[37] Arcos, D; Del Real, RP; Vallet-Regi, M. A novel bioactive and magnetic biphasic material. *J. Biomat.* (2002) 23, 2151-2158.
[38] Hench, LL. Bioceramics. Bioactive Glasses and Glass-Ceramics. *Mater. Sci. Forum* (1999) 293, 37-64.
[39] Ebisawa, Y; Miyaji, F; Kokubo, T; Ohura, K; Nakamura, T. Bioactivity of ferromagnetic glass-ceramics in the system FeO-Fe_2O_3-CaO-SiO_2. *J. Biomat.* (1997) 18, 1277-1284.

[40] LeGeros, RZ; LeGeros, JP. Dense Hydroxyapatite. In: Hench, LL; Wilson, J. editors. *An Intoduction to Bioceramics.* Advanced Series in Ceramics-Vol. 1. Singapore: World Scientific; 1993; 139-180.
[41] Hill, RJ; Howard, CJ. Quantitative Phase Analysis from Neutron Powder Diffraction Data Using the Rietveld Method. *Acta Cryst.* (1987) 20, 467-474.
[42] Bozorth, RM. Ferromagnetism. New York: IEEE Press; 1978.
[43] Craig, DJ. Magnetic Oxides. New York :John Wiley; 1975.
[44] Bozorth RM. Ferromagnetism. New York: IEEE Press; 1978. p. 242.
[45] Banerjee, SK; Moskowitz, BM. Ferrimagnetic properties of magnetite: A New Magnetism. Editors Kirschvink JL et al., Plenum Publishing Corporation, 17-41 (1985).
[46] H. Konaka, F. Miyaji and T. Kokubo: J. Ceramic Soc. Japan Vol. 105 (1997), p.833.
[47] Salinas, AJ; Roman, J; Vallet-Regi, M; Oliveira, JM; Correia; Fernandes, MH. In vitro bioactivity of glass and glass-ceramics of the $3CaO{\cdot}P_2O_5$–$CaO{\cdot}SiO_2$–$CaO{\cdot}MgO{\cdot}2SiO_2$ system. *Biomat.* (2000) 21, 251-257.
[48] Li, R;. Clark, AE; Hench, LL; J. Appl. Biomat. **2**, 231 (1991).

In: Biomaterials Research Advances
Editor: J. B. Kendall, pp. 183-203
ISBN: 978-1-60021-892-7

Chapter 7

STRATEGIES FOR SCAFFOLD VASCULARIZATION IN TISSUE ENGINEERING

Thorsten Walles[1] and Heike Mertsching[2]
[1]Department of Thoracic Surgery, Schillerhöhe Hospital, Stuttgart-Gerlingen, Germany
[2]Cell Systems, Fraunhofer Institute for Interfacial Engineering and Biotechnology (IGB), Stuttgart, Germany

ABSTRACT

Tissue engineering represents a biology driven approach by which biological tissues are engineered through combining material technology and biotechnology. Its advantage over other tissue replacement techniques are several, *e.g.* use of autologous cells, nonimmunogenecity, no side-effects related to foreign graft materials, and potential to grow when implanted into children. Autologous cells of the tissue recipient are seeded on matrices that are fashioned from natural materials, or from synthetic polymers. The cell-matrix constructs are cultured *in vitro* to constitute a bioartificial tissue, the engineered implant for reconstructive surgery. In *in vitro* applications, bioartificial tissues serve as test systems for pharmaceutical drug screening and patient specific therapy. However, tissue engineering of complex tissues and organs is limited by their need of a vascular supply to guaranty graft survival and render bioartificial organ function. Therefore numerous strategies have been developed to overcome this hurdle including indirect revascularization, the concept of wrapping the generated graft with viable tissue, and stimulating ingrowth of microvessels by angiogenic factors, cells and stem cells. The development of a primary vascularized biological scaffold providing a vascular tree including a capillary network for the engineered implant may afford vascular anastomosis of any bioartificial construct to the recipient blood supply.

ABBREVIATIONS AND ACRONYMS

aFGF	acidic fibroblast growth factor
Ang	angiopoetin
bFGF	basic fibroblast growth factor
BioCaM	biological capillarized matrix
EC	endothelial cell
ECM	extracellular matrix
EPC	endothelial precursor cells
FGF	fibroblast growth factor
HDMEC	human dermal microvascular endothelial cells
HDMSC	human bone marrow derived mesenchymal stem cells
hESC	human embryonic stem cells
HGF	hepatocyte growth factor
HUVEC	human umbilical vein endothelial cells
PDGF	platelet-derived growth factor
PERV	porcine endogenous retrovirus
SFF	solid free-form fabrication
TE	tissue engineering
TGF-ß	transforming growth factor beta
VEGF	vascular endothelial cell growth factor

1. APPLIED MATERIALS IN TISSUE ENGINEERING

Tissue engineering (TE) was originally developed as an alternative therapy for the treatment of tissue loss or end-stage organ failure resolving the shortage in tissues and organs for transplantation therapy [1, 2]. In many surgical subspecialities, therapeutic options for surgical reconstruction of organ defects are limited and, regrettably, no dependable synthetic or biologic reconstruction materials have yet been found. Here, TE, the generation of surgical replacement structures from biodegradable carrier structures and autologous cells might offer realistic alternatives [1].

TE is an interdisciplinary field that applies the principles of engineering and the life sciences toward the development of biological substitutes that restore, maintain, or improve tissue function [3]. It offers the potential to create replacement structures from biodegradable scaffolds and autologous cells [1]. The process of fabricating new, physiologic, functioning tissues may be obtained by

1. **guided tissue regeneration** with engineered matrices,
2. injection of autologous or allogenic or xenogenic cells (**cell therapy**), or
3. use of cells seeded on or within matrices (**cell matrix construct**),

with the latter two approaches being the most common.

The use of isolated cells avoids potential surgical complications and allows cell manipulation before injection but holds the risk of possible rejection or loss of cell function [4]. The use of seeded matrices represents the most common approach in TE and requires matrices that are biocompatible, bioabsorbable, nonimmunogenic, supportive of cell attachment and growth, and induce angiogenesis (Tab. 1). These matrices give bioartificial implants their macroscopic threedimensional shape and function as carrier structures for cells, establishing a threedimensional framework for cell seeding. Besides their role in providing mechanical support they participate in cellular communication pathways. Porosity, surface chemistry, topography, threedimensional architecture, immungenecity and mechanic parameters are matrix properties. Cell matrix constructs may be either created by isolating the cells from the host's body with permeable membrane allowing exchange of nutrients (closed system) or by culturing isolated cells *in vitro* and seeding them onto a scaffold, either synthetic or natural, that is implanted into the host after a given cultivation time [1].

Cellular growth and differentiation can be influenced by the (bio)chemical and ultrastructural composition of the matrix, the so called microenvironment [5, 6]. Any synthetical and biological material may serve as matrix for TE.

Matrices can, in theory, be acellular implanted, confining on cellular ingrowth and *in vivo* tissue formation – a concept referred to as **guided tissue regeneration** [7]. Experimental and clinical evidence, however, recently showed that these matrices undergo a fulminant interfibrillar calcification process and do not result in functional bioartificial tissues [8, 9]. Hoerstrup and colleagues showed, that bioartificial implants need an adequate generation of extracellular matrix to guarantee long-time stability and function [10].

To meet the present and future biomaterials challenges successfully, materials scientists and engineers are needed who are familiar with and sensitive to cellular, biochemical, molecular and genetic issues and who work effectively in teams of professionals who include molecular biologists, biochemists, geneticists, physicians and surgeons [11].

Synthetic matrices

Ample experience exists regarding the use and biocompatibility of synthetic implants in reconstructive surgery. For decades they have been applied mainly in dentistry and orthopedic- and cardiovascular surgery. Synthetic materials currently used for biomedical applications include metals and alloys, polymers and ceramics.

Metals have been used almost exclusively for load bearing implants (f.e. joint replacements in orthopedic surgery). Three material groups dominate biomedical metals: 316L stainless steel, cobalt-chromium-molybdenum alloy, and pure titanium alloys. **Ceramics and glasses** have been used less extensively than either metals or polymers. The major drawbacks to their use as implants are their brittleness and poor tensile properties. **Polymers** are the most widely used materials in biomedical applications. They are the materials of choice for cardiovascular devices as well as for replacement and augmentation of various soft tissues (for a detailed review please refer to [11]). Virtually all synthetic matrices for TE applications are generated from polymeric materials.

Polymers are organic materials consisting of large macromolecules composed of many repeating units. Additionally, polymers may contain various (often unspecified) additives, traces of catalysts, inhibitors, and other chemical compounds needed for their synthesis. Over time in the physiological environment, these compounds can leach from the polymer surface. The chemicals released from polymers (f.e. ethylene oxide) may induce adverse local and systemic host reactions that cause clinical complications.

Certain polymers have been designed to undergo controlled degradation. Among biodegradable polymers, poly(lactic acid) (PLLA), poly(glycolic acid) (PGA), and their copolymers have been the most widely used. They facilitate the control of material properties such as strength, degradation time, porosity, and microstructure [11]. Cell attachment can be improved by modifying the polymer chemically or by coating it [12]. With continued improvements in the design of biodegradable polymer matrices, it is now possible to maintain transplanted cells in a more hospitable physiological environment and thus enhance new tissue growth and organization [13]. However, polymer matrices show reduced extracellular matrix (ECM) remodeling combined with a minor growth capability [14].

For the most part, polymers in current biomedical applications do not elicit significant immunologic responses. Design of biodegradable polymers for temporary residence in the body as well as for TE applications has focused on developing materials that degrade into byproducts (such as lactic, glycolic, and caproic acids) that are either biocompatible or are already present in the body, and thus can be either excreted or removed via existing metabolic pathways [11].

Bioartificial implants generated from synthetic matrices have to withstand the established clinical and industrial proceedings to sterilize implants for biomedical applications by moist heat and high pressures (typical conditions in steam auoclaves), ethylene oxide gas, and gamma radiation.

Biological matrices

Tissues consist of pertinent and specific cells and of ECM that is formed and maintained by chemical compounds synthesized by cells. The ECM is a dynamic milieu. It is the locus of growth factor accumulation and it provides the substrate that anchorage-dependent cells need to firmly adhere; this event is prerequisite for subsequent cell functions. It also serves as a store for various growth factors and proenzymes involved in vessel development. Collagen is the most abundant protein found in higher vertebrates. Nineteen types of collagen have been identified. Type I is the most abundant. The ECM contains noncollagenous adhesive proteins that play important roles in organizing the matrix and in enabling cells to attach to it. The most studied of these proteins is fibronectin [11].

Biological matrices are 1) composed artificially from single or few ECM components, such as collagen and fibronectin or 2) generated from threedimensional biologic auto-, allo- or xenogenic tissues. *In vitro* manufacturing allows the combination of ECM proteins and various additives as i.e. growth factors or RGD-peptides to optimize cellular microenvironment and graft stability. Synthesized biological scaffolds usually miss the natural threedimensional structure and the influence of many ECM proteins, changing the microenvironment for ingrowing cells.

Recently, mouldless manufacturing techniques, known as solid free-form fabrication (SFF), or rapid prototyping, have been successfully used to fabricate complex scaffolds. Similarly, to achieve simultaneous addition of cells during the scaffold fabrication, novel robotic assembly and automated threedimensional cell encapsulation techniques are being developed. As a result of these technologies, tissue-engineered constructs can be prepared that contain a controlled spatial distribution of cells and growth factors, as well as engineered gradients of scaffold materials with a predicted microstructure [15].

Electrospinning is a fabrication process that uses an electric field to control the deposition of polymer fibers onto a target substrate. This electrostatic processing strategy can be used to fabricate fibrous polymer mats composed of fiber diameters ranging from several microns down to 100 nm or less. Electrospinning is a rapid and efficient process that can be used to selectively deposit polymers in a random fashion or along a predetermined and defined axis. Thus it is possible to tailor subtle mechanical properties into a matrix by controlling fiber orientation. Electospun collagen promotes cell growth and the penetration of cells into the engineered matrix [16].

Native xeno- or allogenic tissues are decellularized *in vitro* to maintain whole natural ECM composition and are an alternative to biodegradable synthetic scaffolds. Natural xenogenic scaffolds have been used successfully by several groups for pre-clinical tissue engineering, i.e. heart valves, blood vessels and urinary bladder [17]. Concerns regarding the possibility of porcine endogenous retrovirus (PERV) transmission [18-20] onto human recipients by acellular xenogenic matrices have been disproved by experimental and clinical findings [21-23]. However, remaining allo- or xenogenic cellularization would trigger graft rejection [9, 24].

2. Graft Vascularization

The advances in cell and tissue culture techniques and the progress in materials research led to the development of uncounted bioartificial tissues for experimental and clinical applications: Skin-[25, 26] and cartilage [27, 28] transplants generated from autologous tissue biopsies are already commercially available and applied clinically in reconstructive surgery. Heart valves [29], blood vessels [10, 30] and bone [31] are currently tested in clinical studies. Numerous bioartificial repair tissues like urinary bladder [32], urethra [33], and neural tissue [34] are tested experimentally for potential future human application. The multitudinous transplantation studies of the past displayed the elementary importance of tissue vascularization for graft function. Interestingly, the (clinically) realized TE applications represent simple tissues that are supplied by diffusion, but not by vascularization. Therefore, one of the major obstacles in TE of thick, complex tissues is to keep the construct viable *in vitro* (during cultivation and formation of the tissue) as well as *in vivo* (on implantation) [35]. *In vivo*, the construct must be vascularized immediately to allow for its survival and later integration since the host's vascularization is not sufficient to feed the implant [35, 36]. The importance of this issue led to the development of various strategies and experimental approaches to vascularize TE grafts directly or indirectly [37]. None the less, studies on bioengineered tissue vascularization have not been performed until recently (Fig 1).

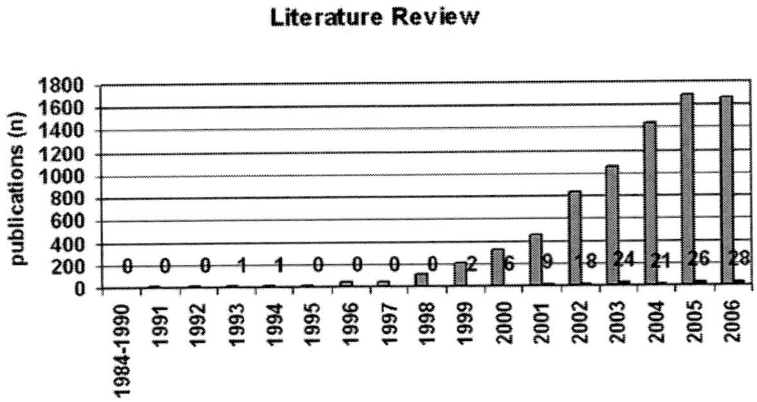

Figure 1: General survey about the literature published in the tissue engineering field. Medline research using the terms "tissue engineering" (grey columns) and "tissue engineering and vascularization" (black columns and numbers). Tissue engineering has emerged as a self-contained scientific field with a growing body of literature in the last decade. The fundamental role of bioartificial tissue vascularization has not been addressed until recently.

2.1 Indirect revascularization

When *in vitro* engineered cellular constructs are transplanted *in vivo* they have to rely on processes like interstitial fluid diffusion and blood perfusion. Here recites a core limitation for transfer of tissue engineering models from the *in vitro* to the *in vivo* environment [38]. Diffusion is the initial process involved, but it can only provide for cell support within a maximum range of 200 μm into the matrix [39, 40]. Therefore, prompt formation of new blood vessels by budding or sprouting of preexisting vessels is necessary to sustain the newly formed tissue. Animal experiments investigating the capillary ingrowth in a rodent transplant model showed that avascular tissues with a diameter of less than 1 mm are supplied by diffusion over a time period of 2 weeks [36]. Tissue revascularization was detectable about 2 weeks following transplantation and continued up to the 8th postoperative week [8].

Indirect revascularization, the concept of wrapping the TE graft with viable tissue (i.e. greater omentum, peritoneum, muscle, subcutaneous tissue) and stimulating ingrowth of microvessels, has been intensively investigated in the experimental literature [37, 41, 42]. In this approach, the new capillaries emerge from neighboring vessels induced by an inflammatory wound healing response as a reaction to the surgical implantation [43]. This, combined with the hypoxia within the implant evokes local expression of angiogenic growth factors. However, it has been shown that grafts implanted into i.e. the peritoneal cavity are covered with a heavy vascularized epithelial layer, representing a foreign-body reaction encapsulating the implanted graft. This reaction does not per se promote neoangiogenesis in the graft [44]. Therefore it is highly questionable if TE implants transplanted into the recipient will be viable following this approach.

2.2 Induction of angiogenesis *in vitro*

The vascular network branches in a hierarchical fashion and is organized spatially to provide adequate nutrients to the cells of all organs and supporting structures by diffusion and convection. The maturation of nascent vasculature requires recruitment of mural cells, generation of an ECM and specialization of the vessel wall for structural support and regulation of vessel function. In addition, the vascular network must be organized so that the parenchymal cells receive adequate nutrients. All of the processes are orchestrated by physical forces as well as by a constellation of ligands and receptors whose spatio-temporal patterns of expression and concentration are tightly regulated [45].

Physicists, chemists, engineers, material scientists, biologists and physicians try to mimic the only fragmentary understood biological processes of vasculogenesis, angiogenesis and vessel maturation. Three approaches have been used for vascularization of bioengineered tissues:

1. incorporation of angiogenic factors in the bioengineered tissue
2. seeding endothelial cells (EC) with other cell types
3. prevascularization of matrices prior to cell seeding.

Optimized surgical implantation techniques are one more essential step towards clinical application of engineered bioartificial tissues [38].

2.2.1 Angiogenic factors

Angiogenesis is a complex process that relies on the presence of an ECM as well as on migration and mitogenic stimulation of endothelial cells (EC). Various growth factors (primarily fibroblast growth factor (FGF) and vascular endothelial cell growth factor (VEGF)), released by macrophages and EC in the hypoxic and acidic environment of the implant site, stimulate growth of new capillaries into the neotissues, thus supplying the life-sustaining oxygen and nutrients contained in circulating blood [45-48]. Neovascularization involves proliferation, maturation, and organization of EC into capillary tubes.

2.2.1.1 Direct angiogenic factors

The vascular endothelial growth factor (VEGF) represents the key participant in these complex well orchestrated processes and is highly responsible to environmental stimuli such as hypoxia, low pH, abnormal hydrostatic pressure and shear stress [49]. As a specific mitogen of ECs, VEGF promotes differentiation and proliferation of EC leading to the formation of immature vessels [45].

Basic fibroblast growth factor (bFGF) acts as a strong fibroblast mitogen, and is also a highly angiogenic chemotactic factor for EC. bFGF and acidic FGF (aFGF) are localized to the heparan-sulfate-proteoglykane (HPSG) on the cell surface and in the ECM. Besides their angiogenic activity, the FGFs are important for wound healing and tissue repair [50].

Hepatocyte growth factor (HGF) participates in several biological processes such as embryogenesis, organ regeneration, wound healing, and angiogenesis. HGF regulates migration and proliferation of human microvascular EC *in vitro* and induces the formation of new blood vessels *in vivo* [50].

2.2.1.2 Indirect angiogenic factors

A large group of angiogenic growth factors that do not induce EC proliferation or migration *in vitro*, were shown to induce angiogenesis *in vivo*. These factors are considered indirect angiogenic factors [50].

PDGF is a potent stimulator of growth and motility of fibroblasts and smooth muscle cells, but also acts on EC and neurons. PDGF induces an angiogenic response *in vivo* but does not affect EC proliferation directly. PDGF supports the formation of functional vascularized connective tissue in wound healing during tissue repair. PDGF receptors are found only on a subset of EC, mostly on developing endothelial tubes and on microvascular EC, but not on quiescent EC. It is not clear yet whether PDGF can initiate angiogenesis *in vitro* or support an ongoing angiogenic process.

Transforming growth factor-ß (TGF-ß) inhibits EC proliferation *in vitro* but promotes angiogenesis *in vivo*. TGF-ß and its receptors are important regulators of EC differentiation and establishment and maintenance of vessel wall integrity. It seems that TGF-ß has a different function on EC and vessel formation at different stages of the angiogenic process.

The angiopoetins Ang 1 and Ang 2 do not seem to enhance EC proliferation, but they have an important role in angiogenesis *in vivo* [50].

The use of angiogenic factors is a popular approach to induce neovascularization in engineered tissues (Table 1). However, bolus injection of recombinant angiogenic proteins such as VEGF and bFGF has limitations due to the inherent instability of these proteins *in vivo* and the possibility of uncontrolled effects at distant sites [51, 52].

Table 1 - Properties of the ideal tissue engineering matrix:

Biocompatibility
Bioabsorbability
Non-immunogenicity
Support cell attachment and growth
Induction of angiogenesis
Threedimensional structure
Affording sterilization

2.2.2 Cells

Endothelial cells act as the main mediator of neovascular growth. They are forming the endothelial layer of the vasculature or exist as circulating EC in the blood stream. During angiogenesis a directed migration of EC occurs. For the elongation of the developing blood vessel EC proliferation is necessary. The stabilization of the newly formed vessel is achieved by the recruitment of further cell types and the production of ECM compounds [45, 58].

EC have been used for generation of capillary-like structures and vessels *in vitro* by different groups [54, 57, 72] (Table 2). *In vivo* these cells are supposed to form networks of capillaries and gain access to the recipient's circulation [73]. The newly formed vascular networks display a differentiated morphology. However, the processes that occur following implantation are not fully understood so far and there are only scarce *in vivo* data on the efficiency of this approach with regard to enhance and accelerate vascularization of bioartificial tissues [38, 74].

It has been shown that the use of primary cells in TE has certain limitations: Primary EC not only have a limited lifespan and reduced phenotype expression with each passage *in vitro*, but contaminating cells of other types may also be present. Furthermore, EC of different tissue origins are highly heterogenous and microvascular EC are the relevant cells involved in vascularization and angiogenesis [50].

To overcome these limitations, endothelial progenitor cells have been applied. ECs can be derived from embryonic stem cells (ESC) [70] or progenitor cells [75, 86]. The application of these cells faces the problems of controlling differentiation and avoiding teratogenesis. Immunological concerns regarding the use of hESC-derived endothelial cells hold back their potential clinical application [77].

Since cell-containing constructs require immediate supply with nutrients and oxygen after implantation and even with synchronous transplantation of EC some time is needed for formation of anastomosis between newly formed capillaries and the recipient's circulation, the potential benefits of EC transplantation for induction of vascularization are questionable [78].

Table 2: Angiogenic factors and cells applied to induce tissue vascularization in cell-matrix constructs following implantation*.

Angiogenic factors	*References*	Cells	References
ECGF	*53*	primary EC	*56, 57, 67, 72*
VEGF	*54*	EC peripheral blood	*56*
bFGF	*54*	human EC lines	*56, 58*
TGF beta 1	*55*	HUVEC	*54, 56, 57, 59*
		HDMEC	*56, 57, 60*
		stem cells	*51, 62*
		HDMSC	*53, 64, 68, 69*
		hESC	*65, 70, 71*
		HPMEC	*30, 57*
		EPC	*33, 58, 66*

* selected references, list is not exhaustive

2.2.3 Strategies

It has been realized by many groups that the solely application of single cell types or growth factors does not result in vascularized bioartificial tissues. Therefore they developed strategies to combine a various number of angiogenic stimuli in the generation process of their tissues (Table 3).

2.2.3.1 Culture conditions

Biomechanical stimuli direct the process of EC differentiation *in vitro* and *in vivo* [72, 79]. Therefore a pulsatile hemodynamic flow pattern mimicking physiological conditions is essential for the differentiation of EC and their function as endothelial barrier between bloodstream and tissue. It was shown, that vascular grafts generated from EC and mural cells that were cultered under pulsatile conditions lead to functional arteries when implanted in miniature swine [72].

Table 3: Strategies to improve the angiogenesis potential of TE scaffolds*.

Strategy	References
Sustained growth factor release by acellular biodegradable polymer	
Fibrin gels + growth factors	*80*
Heparin binding release systems + growth factors	*81, 95, 96, 99*
Encapsulation of bFGF	*97, 98*
Heparin + collagen + VEGF	*85*
Injectable collagen + TGF-ß	*55*
Chondroitin sulfate + bFGF	*100*
Degradable polymer scaffold + VEGF	*82*
Degradable polymer scaffold + encapsulation of VEGF + PDGF	*51*
Scaffold cellularization prior to implantation	
Degradable polymer scaffold + EC	*101*
Degradable polymer scaffold + growth factor + EC	*102, 103*
Degradable polymer scaffold + EPC	*75*
Genetical modification of cellularized scaffolds	
Degradable polymer scaffold + plasmid PDGF	*83*
Degradable polymer scaffold + VEGF transfected cells	*50, 89*
Degradable polymer scaffold + VEGF transfected cells + growth factors	*104*
Degradable polymer scaffold + telomerized cells	*90, 91*

* selected references, list is not exhaustive

2.2.3.2 Scaffold modifications

Angiogenic factors were incorporated into bioengineered tissues prior to implantation in order to enhance new capillary ingrowth from the host vascular network [43]. Immobilization of angiogenic growth factors, for instance in fibrin gels or by using heparine-binding release systems, allows for optimized release kinetics and longer lasting effects [80, 81]. In an attempt to achieve low levels and continuous release of VEGF *in vivo*, porous degradable polymer scaffolds containing VEGF were fabricated by including the growth factor in a gas foaming. Sustained release of VEGF was achieved for at least 15 days [82]. PLG matrices were loaded with plasmid, which was subsequently released over a period ranging from days to a month *in vitro*. Sustained delivery of plasmid DNA encoding PDGF from the matrices led to transfection of large numbers of cells, resulting in enhanced matrix deposition and blood vessel formation in the developing tissue [83].

The first systems were designed to release only one growth factor, but vascular formation requires several types of angiogenic growth factors at different times during development. First polymeric systems that allow the tissue-specific delivery of two or more growth factors, with controlled dose and rate of delivery, were recently introduced [51].

While a variety of highly innovative matrices are under *in vitro* and *in vivo* evaluation, less sophisticated clinically established and approved biomaterials are readily available for first applications [50, 84]. The use of polymer matrices preloaded with growth factors, matrix molecules, or vascular cells have already proven their usefulness in developing strategies designed to enhance EC growth, survival, and vessel assembly [46, 51, 84].

Nanofabrication techniques are currently under way to engineer network structures that will mimic the capillary network, expanding from and merging back to a single major vessel [77, 85, 86]. Ink-jet printing can be used to pattern cells into tubular structures [87].

2.2.3.3 Co-cultures

Endothelial cells or endothelial precursor cells (EPC) may be incorporated into the bioengineered tissue to enhance tissue revascularization. They may serve as "short-term delivery devices" for angiogenic stimuli by overexpressing and secreting soluble angiogenic factors to enhance the initial vascularization of engineered tissues [13, 50, 88]. Various types of EC such as human umbilical vein EC (HUVEC), and human dermal microvascular EC (HDMEC) were used for *in vitro* models of angiogenesis and vasculogenesis (Table 2). Gel-like matrices such as fibrin have been used for cell immobilization [80, 81]. Cultured EC rapidly underwent apoptosis following implantation *in vivo*. A combination of EC seeding with the insertion of a constitutive source of angiogenic factor secretion may be advantegous [46].

2.2.3.4 Cell modifications

Cells within the bioengineered tissue can be genetically altered to secrete recombinant angiogenic factors by naked (plasmid) DNA and viral vectors. To preserve EC in bioengineered tissue *in vivo*, cells were infected with adenovirus vectors that encode for VEGF [50, 89]. Human EC transduced with human telomerase reverse transcriptase (hTERT) are capable of forming vascular structures in subcutaneous implants. The telomerized EC form a stable, more long-lived vasculature compared to normal EC [90, 91].

Gene transfer strategies may provide for efficient stimulation of vascularization within bioartificial tissues, but further insight into long-term effects, phenotypical stability and application techniques need to precede any wide-spread application in humans [46].

2.2.3.5 Tissue prevascularization

The prevascularization of the supporting polymer prior to or even after cell seeding represents a multi-stage surgical approach to generate bioartificial tissue vascularization [38]. First, a preformed tissue or acellular matrix [92] is implanted into a region with a good vascular supply (f.e. muscle) suitable for later microsurgical transfer. The bioengineered tissue is organized around the vascular network, providing tissue perfusion. In a second operation the autologous implant is harvested en-bloc with the surrounding tissue preserving its vascular pedicle as a free flap. The implant is transplanted into the defect site and its vascular pedicle is connected to the local circulation by means of a microvascular anastomosis. Prefabricated bioartificial tissue grafts have been clinically applied in different settings [38, 93, 94].

2.3 Biological capillarized matrix (BioCaM)

The angiogenic capabilities of most engineered tissues are still not sufficient. There is thus a need for materials in which the angiogenesis keeps pace with the ingrowth of invading cells. The generation of a viable and functional TE implant can, in the long run, only be reached with bioartificial grafts enduing a fully developed capillary network connected to the recipient circulation [105] (Fig. 2). ECs generate an active antithrombotic surface that facilitates transit of plasma and cellular constituents. For proper function of the endothelium, integrity is required, i.e. mainly ensured by endothelial cell-to-cell junctions. A functional vascular endothelial lining is of paramount importance to avoid graft thrombosis and failure [106]. Recently, a unique biological capillarized matrix (BioCaM) was introduced that supplies a capillary bed for the nutrition of complex man-made biological implants [65, 107]. The matrix represents an acellular collagenous network generated from porcine tissue by removal of all cellular components, similar to the SIS matrix that has been characterized recently as a complete absorbable biocompatible framework [108], enabling cellular migration and differentiation [109, 110], showing early capillary ingrowth and endothelialization as well as high infection resistance [111]. The SIS is widely used in reconstructive surgery. Implanted SIS samples induced no rejection in autogene, allogene, and xenogene transplantation models [108, 109]. The SIS was successfully used as a biological matrix for numerous body tissues structures. The applicability has been shown in animal studies and in human clinical trials including musculoskeletal, cardiovascular [112, 113] urogenital [114, 115], medial collateral ligament [116] and for laryngeal reconstruction [117]. The ECM composition of the SIS provides a complete and fast remodeling after implantation [114], which is an important prerequisite for tissue maturation [118, 119] and physiological function of the implant. The applicability of the SIS matrix and similar biological or synthetical scaffolds for tissue engineering, however, is limited by the lack of graft vascularization. In the BioCaM matrix the vascular structures are restored *in vitro* by autologous endothelial precursor cells. The BioCaM composition allows its application as a matrix for a wide variety of tissue engineering applications and may overcome the hurdle of lacking bioartificial graft vascularization [65].

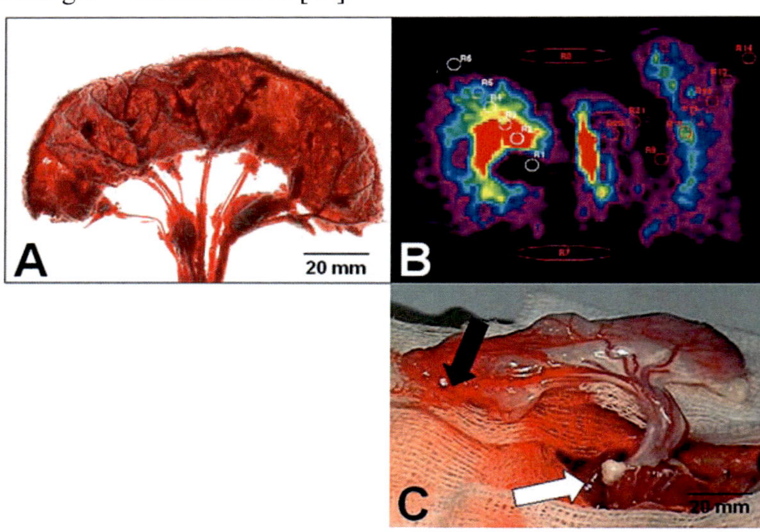

Figure 2: Biological capillarized matrix (BioCaM) for tissue engineering of complex organ structures. **A** A decellularized porcine small bowl segment serves as scaffold matrix for bioartificial tissue generation. **B** Thin layer PET scanning illustrating the metabolic activity in reseeded scaffolds (middle and left) and acellular controls (right) (red = high metabolic activity; blue = no metabolic activity). **C** The reseeded vascular network can be connected to the recipient blood supply by an arteriovenous anastomosis (white arrow). The bioartificial tissue is perfused entirely including the peripheral capillary bed (black arrow).

2.3.1 Clinical application

Clinically, the BioCaM matrix has been applied successfully to generate a bioartificial airway patch in one patient with an extensive tracheobronchial insufficiency [118]. The tissue was incorporated into the native trachea without signs of infection, rejection or necrosis. The luminal patch surface was reseeded with functional respiratory epithelium. There was no evidence of granulation tissue formation at the implantation site, but surprisingly an extensive tissue maturation process following implantation [74].

Those anecdotal clinical reports denote first tissue engineering applications entering medical practice. Currently it is still unknown, if these new types of implants will tolerate the specific needs in cancer patients undergoing postoperative chemo- and radiotherapy. Applied in oncologic patients, eventually necessitating postoperative radiotherapy for tumor treatment, those biological tissues would have to endure high irradiation in the early postoperative phase without graft degeneration. Biancosino et *al.* subjected bioartificial patch tissue experimentally to radiation treatments mimicking clinically established radiation protocols. It was shown that the bioartificial tissue devitalizes following radiation with 45 Gy but recovers within 6 weeks. These findings illustrate the applicability of bioartificial implants for surgical reconstruction in oncologic patients [120].

2.3.2 Bioartificial tissues as model systems

In addition to clinical applications, TE can provide new tools for studying cell and developmental biology as well as for better understanding angiogenesis processes by providing approaches for cell and tissue growth in threedimensional *in vitro* environments [35].

Like any other healthy tissue, tumor tissue depends on vascularization and adequate supply of oxygen and nutrients. Angiogenesis plays an important role in tumor growth and tumor spread. Tumor cells release cytokines and proteins that induce angiogenesis to produce a supporting tumor-vasculature. Ultimately, this results in a disproportionate vascular supply of tumor tissues [45]. These are the fundamentals for rasant and unbounded tumor growth and tumor spread. Anti-angiogenesis is the therapeutic approach to interrupt the mechanisms of angiogenesis in growing tumor tissue: to uncouple the growing tumor tissue from its vascular supply to inhibit metastasis formation and to induce necrosis of the tumor tissue [46]. The development of anti-angiogenetic strategies centers efforts in the pharmaceutical industry. However, results from the animal models available for studying angiogenesis and anti-angiogenesis are often not transferable into man [120]. New threedimensional test-systems applying human cells or tissues are demanded by physicians, scientists and the industry []. Vascularized bioartificial human tumor and mosaic tissues (composed of healthy and tumor tissue) could, at least in part, bridge this gap: Those vascularized artificial tissues could be investigated and treated *in vitro*: they allow testing new compounds or drugs outside a patient, but still inside a human tissue, before they are carried into clinical evaluation. Moreover, numerous cost- and time-intensive animal experiments could become superfluous. This could ideally results in shortened research and development periods for drugs. This time saving would cut R&E expenses.

3. Summary

Tissue engineering offers potential applications in almost all surgical subspecialities for generating tissue and organ replacements. Beyond the clinic, tissue engineering applications are interesting as humanoid test-systems in basic and applied research. With the emergence of tissue engineering as a discipline, it has become increasingly clear that long-term success in organ and tissue reconstruction will depend on the ability to develop a stable, renewable supply of blood vessels. The tissue engineering field profits from the growing and fast developing understanding of angiogenesis, the appreciation and perception about vascular formation by vasculogenesis and angiogenesis . At present, first steps towards the realization of vascularized bioartificial tissues have been undertaken. Diverse concepts regarding choice of matrices, cell types, culture conditions and strategies have been followed. In the figurative sense, mosaic pieces have been found. To assemble those pieces, a deeper and more sound understanding of cell-cell and cell-matrix interactions is needed.

References

[1] Fuchs, JR; Nasseri, BA; Vacanti, JP. Tissue engineering: A 21st century solution to surgical reconstruction. *Ann Thorac Surg* 2001; *72*: 577-591.

[2] Atala, A. Tissue engineering and regenerative medicine: concepts for clinical application. *Rejuvenation Res* 2004; *7*: 15-31.

[3] Langer, R; Vacanti, JP. Tissue engineering. *Science* 1993; *260*: 920-6.

[4] Vacanti, JP; Langer, R. Tissue engineering: The design and fabrication of living replacement devices for surgical reconstruction and transplantation. *Lancet* 1999; *354* (Suppl): S132-4.

[5] Abbott A. Biology's new dimension. *Nature* 2003; *424*: 870-2.

[6] Krieg, T; Le Roy, EC. Diseases of the extracellular matrix. *J. Mol. Med.* 1998; *76*: 224-225.

[7] Skalak, R; Fox, CF. Tissue engineering. *Ann. Biomed. Engl* 1991; *19*: 529-533.

[8] Walles, T; Puschmann, C; Haverich, A; Mertsching, H. Acellular scaffold implantation – no alternative to tissue engineering. *Int. J. Art. Org.* 2003; *26*: 225-34.

[9] Simon, P; Kasimir, MT; Seebacher, G; Weigel, G; Ullrich, R; Salzer-Muhar, U; Rieder, E; Wolner, E. Early failure of the tissue engineered porcine heart valve SYNERGRAFT in pediatric patients. *Eur. J. Cardiothorac. Surg.* 2003; *23*: 1002-6.

[10] Hoerstrup, SP; Zund, G; Schnell, AM; Kolb, SA; Visjager, JF; Schoeberlein, A; Turina M. Optimized growth conditions for tissue engineering of human cardiovascular structures. *Int. J. Art. Org.* 2000; *23*: 817-823.

[11] Dee, KC; Puleo, DA; Bizios, R. An introduction to Biomaterial Interactions. 1st Edit.; New Jersey, *John Wiley & Sons, Inc.;* 2002

[12] Corneille, S; Schacht, E. Surface plasmon resonance: a technique for characterizing biomaterial surfaces. *Med Device Technol* 1998; *9(1)*: 28-31.

[13] Putnam, AJ; Mooney, DJ. Tissue engineering using synthetic extracellular matrices. *Nat Med* 1996; *2*: 824-6.

[14] Thompson, MM; Budd, JS; Eady, SL; Allen, KE; James, M; James, RF; Bell, PR. Endothelial cell seeding of damaged native vascular surfaces: prostacyclin production. *Eur J Vasc Surg* 1992; *6(5)*: 487-493.

[15] Hutmacher, DW; Sittinger, M; Risbud, MV. Scaffold-based tissue engineering: rationale for computer-aided design and solid free-form fabrication systems. *Trends Biotechnol* 2004; *22(7)*: 354-62.

[16] Matthews, JA; Wnek, GE; Simpson, DG; Bowlin, GL. Electrospinning of collagen nanofibers. *Biomacromolecules* 2002; *3(2)*: 232-8.

[17] Atala, A. Tissue engineering for bladder substitution. *World J Urol* 2000; *18(5)*: 364-370.

[18] Weiss, RA. Transgenic pigs and virus adaption. *Nature* 1988; 391: 327-8.

[19] Patience, C ; Takeuchi, Y ; Weiss, RA. Infection of human cells by an endogenous retrovirus of pigs. *Nat. Med* 1997; *3*: 282-6.

[20] Wilson, CA; Wong, S; Muller, J; Davidson, CE; Rose, TM; Burd, P. Type C retrovirus released from porcine primary peripheral blood mononuclear cells infects human cells. *J. Virol.* 1998; *72*: 3082 – 3087.

[21] Leyh, RG; Wilhelmi, M; Walles, T; Kallenbach, K; Rebe, P; Oberbeck, A; Herden, T; Haverich, A; Mertsching, H. Acellularized porcine heart valve scaffolds for heart valve tissue engineering and the risk of cross-species transmission of porcine endogenous retrovirus. *J Thorac Cardiovasc Surg* 2003; *126*: 1000-4.

[22] Moza, AK; Mertsching, H; Herden, T; Bader, A; Haverich, A. Heart valves from pigs and the porcine endogenous retrovirus: Experimental and clinical data to assess the probability of porcine endogenous retrovirus infection in human subjects. *J Thorac Cardiovasc Surg* 2001; *121*: 697-701.

[23] Walles, T; Lichtenberg, A; Puschmann, C; Leyh, R; Wilhelmi, M; Kallenbach, K; Haverich, A; Mertsching, H. In vivo model for cross-species porcine endogenous retrovirus transmission using tissue engineered pulmonary arteries. *Eur J Cardiothoracic Surg* 2003; *24(3)*: 358-63.

[24] Rieder, E; Kasimir, MT; Silberhumer, G; Seebacher, G; Wolner, E; Simon, P; Weigel, G. Decellularization protocols of porcine heart valves differ importantly in efficiency of cell removal and susceptibility of the matrix to recellularization with human vascular cells. *J. Thorac. Cardiovasc. Surg.* 2004; *127*: 399-405.

[25] Berking, C; Herly, M. Humans skin reconstruct models: a new application for studies of melanocyte and melanoma biology. *Histol Histopathol* 2001; *16(2)*: 669-674.

[26] Dietrich, C; Schandar, M; Noll, M; Johannes, FJ; Brunner, H; Graeve, T; Rupp, S. In vitro reconstructed human epithelia reveal contribution of Candida albicans EFG1 and CPH1 to adhesion and invasion. *Microbiology* 2002; *148*: 497-506.

[27] Jobanputra, P; Parry, D; Fry-Smith, A; Burls, A. Effectiveness of autologous chondrocyte transplantation for hyaline cartilage defects in knees: a rapid and systematic review. *Health Technol Assess* 2001; *5(11)*: 1-57.

[28] Lindahl, A; Brittberg, M; Peterson L. Cartilage repair with chondrocytes: clinical and cellular aspects. *Novartis Found Symp.* 2003; *249*: 175-86.

[29] Gulbins, H; Goldemund, A; Uhlig, A; Pritisanac, A; Meiser, B; Reihart, B. Implantation of an autologously endthelialized homograft. *J Thorac Cardiovasc Surg* 2003; *126(3)*: 890-1.

[30] Shinoka, T; Shum-Tim, D; Ma, PX; Tanel, RE; Isogai, N; Langer, R; Vacanti, JP; Mayer, JE. Jr. Ceation of viable pulmonary artery autografts through tissue engineering. *J Thorac Cardiovasc Surg* 1998; *115*: 536-46.
[31] Mistry, AS; Mikos, AG. Tissue Engineering Strategies for Bone regeneration. *Adv Biochem Engin/Biotechnol* 2005; *94*: 1-22.
[32] Schultheiss, D; Gabouev, AI; Cebotari, S; Tudorache, I; Walles, T; Schlote, N; Wefer, J; Kaufmann, PM; Haverich, A; Jonas, U; Stief, CG; Mertsching, H. Biological vascularized matrix for bladder tissue engineering: matrix preparation, reseeding technique and short-term implantation in a porcine model. *J Urol* 2005; *173(1)*: 276-80.
[33] Atala, A. Regeneration of urologic tissues and organs. *Adv Biochem Eng Biotechnol.* 2005; *94*: 181-210.
[34] Stocum, DL. Stem Cells in CNS and Cardiac Regeneration. *Adv Biochem Engin/Biotechnol* 2005; *94*: 135-159.
[35] Levenberg, S; & Langer, R. Advances in tissue engineering. *Curr Top Devel Biol* 2004; *61*: 113-128.
[36] Walles, T; Herden, T; Haverich, A; Mertsching, H. Influence of scaffold thickness and scaffold composition on bioartificial graft survival. *Biomaterials* 2003; *24*: 1233-1239.
[37] Klepetko, W; Marta, GM; Wisser, W; Melis, E; Kocher, A; Seebacher, G; Aigner, C; Mazhar, S. Heterotopic tracheal transplantation with omentum wrapping in the abdominal position preserves functional and structural integrity of a human tracheal allograft. *J. Thorac. Cardiovac. Surg.* 2004; *127*: 862-7.
[38] Kneser, U; Schaefer, DJ; Polykandriotis, Horch, RE. Tissue engineering of bone: the reconstructive surgeon's point of view. *J Cell Mol Med* 2006; *10(1)*: 7-19.
[39] Folkman, J; Hochberg, M. Self-regulation of growth in three dimensions. *J Exp Med* 1973; *138*: 745-53.
[40] Goldstein, AS; Juarez, TM; Helmke, CD; Gustin, MC; Mikos, AG. Effect of convection on osteoblastic cell growth and function in biodegradable polymer foam scaffolds. *Biomaterials* 2001; *22*: 1279-88.
[41] Glatz, F; Neumeister, M; Suchy, H; Lyons, S; Damikas, D; Mowlawi, A. A tissue-engineering technique for vascularized laryngotracheal reconstruction. *Arch. Otolaryngol. Head Neck Surg.* 2003; *129*: 201-6.
[42] Walles, T. Bioartificial tracheal grafts: Can tissue engineering keep its promise ? *Expert Rev Med Devices* 2004; *1(2)*: 241-50.
[43] Eiselt, P; Kim, BS; Chacko, B; Isenberg, B; Peters, MC; Greene, KG; Roland, WD; Loebsack, AB; Burg, KJ; Culberson, C; Halberstadt, CR; Holder, WD; Mooney, DJ. Development of technologies aiding large-tissue engineering. *Biotechnol Prog.* 1998; *14*: 134-40.
[44] Cebotari, S; Walles, T; Sorrentino, S; Haverich, A; Mertsching, H. Guided tissue regeneration of vascular grafts in the peritoneal cavity. *Circ. Res.* 2002; *90*: e71.
[45] Jain, RK. Molecular regulation of vessel maturation. *Nat Med* 2003; *9(6)*: 685-93.
[46] Polverini, PJ. Angiogenesis in Health and Disease: Insights into basic mechanisms and therapeutic opportunities. *J Dent Edu* 2002; *66(8)*: 962-75.
[47] Ribatti, D; Vacca, A; Presta M. The discovery of angiogenic factors: A historical review. *Gen Pharmacol* 2000; *35*: 227-31.

[48] Ahrendt, G; Chickering, DE; Ranieri, JP. Angiogenic growth factors: a review for tissue engineering. *Tiss Eng* 1998; *2*: 117-30.

[49] Shweiki, D; Itin, A; Soffer, D; Keshet, E. Vascular endothelial growth factor induced by hypoxia may mediate hyperoxia-iniated angiogenesis. *Nature* 1992; *359*: 843-5.

[50] Nomi, M; Atala, A; De Coppi, P; Soker, S. Principals of neovascularization for tissue engineering. *Mol Asp Med* 2002; *23*: 463-83.

[51] Richardson, TP; Peters, MC; Ennett, AB; Mooney, DJ. Polymeric system for dual growth factor delivery. *Nat Biotechnol* 2001; *19*: 1029-34.

[52] Yancopoulos, GD; Davis, S; Gale, NW; Rudge, JS; Wiegand, SJ; Holash, J. Vascular-specific growth factors and blood vessel formation. *Nature* 2000; *407*: 242-8.

[53] Hunter, SK; Kao, JM; Wang, Y; Benda, JA, Rodgers, VG. Promotion of neovascularization around hollow fiber bioartificial organs using biologically active substances. *ASAIO* 1999; *45(1)*: 37-40.

[54] Wenger, A; Stahl, A; Weber, H; Finkenzeller, G; Augustin, HG; Strak, GB; Kneser, U. Modulation of in vitro angiogenesis in a three-dimensional spheroidal coculture model for bone tissue engineering. *Tissue Eng* 2004; *10(9-10)*: 1536-47.

[55] Bentz, H; Schroeder, JA; Estridge, TD. Improved local delivery of TGF-ß2 by binding to injectable fibrillar collagen via difunctional polyethylene glycol. *J Biomed Mater Res* 1998; *39*: 536-48.

[56] Fuchs, S; Motta, A; Migliaresi, C; Kirckpatrick, CJ. Outgrowth endothelial cells isolated and expanded from human peripheral blood progenitor cells as a potential source of autologous cells for endothelialization of silk fibroin biomaterials. *Biomaterials* 2006; 10 [Epub ahead of print]

[57] Unger, RE; Peters, K; Huang, Q; Funk, A; Paul, D; Kirckpatrick, CJ. Vascularization and gene regulation of human endothelial cells growing on porous polyethersulfone (POS) hollow fiber membranes. *Biomaterials* 2005; *26(17)*: 3641-9.

[58] Kirckpatrick, CJ; Unger, RE; Krump-Konvalinkova, V; Peters, K; Schmidt, H; Kamp, G. Experimental approaches to study vascularization in tissue engineering and biomaterial applications. *J Mater Sci Mater Med* 2003; *14(8)*: 677-81.

[59] Kelm, JM; Diaz-Sanchez-Bustamante, C; Ehler, E; Hoerstrup, SP; Djonov, V; Ittner, L; Fussenegger, M. VEGF profiling and angiogenesis in human microtissues. *J Biotechnol* 2005; *118(2)*: 213-29.

[60] Sahota, PS; Burn, JL; Brown, NJ; MacNeil, S. Approaches to improve angiogenesis in tissue-engineered skin. *Wound Repair Regen* 2004; *12(6)*: 635-42.

[61] Laschke, MW; Harder, Y; Amon, M; Martin, I; Farhadi, J; Ring, A; Torio-Padron, N; Schramm, R; Rucker, M; Junker, D; Ha, JM; Carvalho, C; Heberer, M; Germann, G; Vollmar, B; Menger MD. Angiogenesis in Tissue Engineering: Breathing Life into Constructed Tissue Substitutes. *Tissue Eng.* 2006; Jun 1; [Epub ahead of print]

[62] Zwaginga, JJ; Doevendans, P. Stem cell-derived angiogenic/vasculogenic cells: possible therapies for tissue repair and tissue engineering. *Clin Exp Pharmacol Physiol.* 2003; *30(11)*: 900-8.

[63] Okuyama, H ; Krishnamachary, B ; Zhou, YF ; Nagasawa, H ; Bosch-Marce, M ; Semenza, GL. Expression of vascular endothelial growth factor receptor 1 in bone marrow-derived mesenchymal cells is dependent on hypoxia-inducible factor 1. *J Biol Chem.* 2006; *281(22)*: 15554-63.

[64] Urbich, C; Dimmeler, S. Endothelial progenitor cells functional characterization. *Trends Cardiovasc Med.* 2004; *14(8)*: 318-22.

[65] Battista, S; Guarnirei, D; Borselli, C; Zeppetelli, S; Gerbasio, D; Keene, DR; Ambrosio, L; Netti, PA. The effect of matrix composition of 3D constructs on embryonic stem cell differentiation. *Biomaterials* 2005; *26(31)*: 6194-207.

[66] Kaushal, S; Amiel, GE; Guleserian, KJ; Shapira, OM; Perry, T; Sutherland, FW; Rabkin, E; Moran, AM; Schoen, FJ; Atala, A; Soker, S; Bischoff, J; Mayer, JE. Jr. Functional small-diameter neovessels created using endothelial progenitor cells expanded in vivo. *Nat Med* 2001; *7*: 1035-40.

[67] Poh, M; Boyer, M; Solan, A; Dahl, SL; Pedrotty, D; Banik, SS; McKee, JA; Klinger, RY; Counter, CM; Nilason, LE. Blood vessels engineered from human cells. *Lancet* 2005, *365*: 2122-4.

[68] Hibino, N; Shinoka, T; Matsummura, G; Ikada, Y; Kurosawa, H. The tissue-engineered vascular graft using bone marrow without culture. *J Thorac Cardiovasc Surg* 2005; *129*: 1064-70.

[69] Cho, SW; Park, HJ; Ryu, JH; Kim, SH; Kim, YH; Choi, CY; Lee, MJ; Kim, JS; Jang, IS; Kim, DI; Kim, BS. Vascular patches tissue-engineered with autologous bone-marrow derived cells and decellularized tissue matrices. *Biomaterials* 2005; *26*: 1915-24.

[70] Levenberg, S; Golub, JS; Amit, M; Itskovitz-Eldor, J; Langer, R. Endothelial cells derived from human embryonic stem cells. *Proc Natl Acad Sci USA* 2002; *99*: 4391-4396.

[71] Levenberg, S; Huang, NF; Lavik, E; Rogers, AB; Iskovitz-Eldor, J; Langer, R. Differentiation of human embryonic stem cells on three-dimensional polymer scaffolds. *Proc Natl Acad Sci USA* 2003; *100*: 12471-6.

[72] Niklason, LE; Gao, J; Abbott, WM; Hirschi, KK; Houser, S; Marini, R; Langer, R. *Functional arteries grown in vitro. Science* 1999; *284*: 489-93.

[73] Tremblay, PL; Hudon, V; Berthod, F; Germain, L; Auger, FA. Inoculation of tissue-engineered capillaries with the host's vasculature in reconstructed skin transplanted on mice. *Am J Transplant* 2005; *5(5)*: 1002-10.

[74] Walles, T; Biancosino, C; Zardo, P; Macchiarini, P; Gottlieb, J; Mertsching, H. Tissue remodelling in a bioartificial fibromuscular patch following transplantation in human. *Transplantation* 2005; *80(2)*: 284-5.

[75] Yamashita, J; Itoh, H; Hirashima, M; Ogawa, M; Nishikawa, S; Yurugi, T; Naito, M; Nakao, K; Nishikawa, S. Flk1-positive cells derived from embryonic stem cells serve as vascular progenitors. *Nature* 2000; *408*: 92-6.

[76] Yurugi-Kobayashi, T; et al. Effective contribution of transplanted vascular progenitor cells derived from embryonic stem cells to adult neovascularization in proper differentiation stage. *Blood* 2003; *101*: 2675-8.

[77] Levenberg, S. Engineering blood vessels from stem cells: recent advances and applications. Curr Opin Biotechnol 2005; *16*: 516-23.

[78] Mooney, DJ; Mikos, AG. Growing new organs. *Sci Am* 1999; 4: 60-5.

[79] Walles, T; Gorler, H; Puschmann, C; Mertsching, H. Functional neointima characterization of vascular prostheses in human. *Ann Thorac Surg* 2004; *77(3)*: 864-8.

[80] Wong, C; Inman, E; Spaethe, R; Helgerson, S. Fibrin-based biomaterials to deliver human growth factors. *Thromb Haemost* 2003; *89*: 573-82.

[81] Sakiyama-Elbert, SE; Hubbell, JA. Development of fibrin derivatives for controlled release of heparin-binding growth factors. *J Control Release* 2000; *65*: 389-402.

[82] Murphy, WL; Peters, MC; Kohn, DH; Mooney, DJ. Sustained release of vascular endothelial growth factor from mineralized poly(lactide-co-glycoside) scaffolds for tissue engineering. *Biomaterials* 2000; *21*: 2521-7.

[83] Shea, LD; Smiley, E; Bonadio, J; Mooney, DJ. DNA delivery from polymer matrices for tissue engineering. *Nat Biotechnol* 1999; *17*: 551-4.

[84] Yao, C; Prevel, P; Koch, S; Schenck, P; Noah, EM; Pallua, N; Steffens, G. Modification of collagen matrices for enhancing angiogenesis. *Cells Tissues Organs* 2004; *178 (4)*: 189-96.

[85] Fidkowski, C; Kaazempur-Mofrad, MR; Borenstein, J; Vacanti, JP; Langer R, Wang Y. Endothelialized microvasculature based on a biodegradable elastomer. *Tissue Eng* 2005; *11*: 302-9.

[86] Kaihara, S; Borenstein, J; Koka, R; Lalan, S; Ochoa, ER; Ravens, M; Pien, H; Cunningham, B; Vacanti, JP. Silicon micromachining to tissue engineer branched vascular channels for liver fabrication. *Tissue Eng* 2000; *6*: 105-17.

[87] Xu, T; Jin, J; Gregory, C; Hickman, JJ; Boland, T. Inkjet printing of viable mammalian cells. *Biomaterials* 2005; *26*: 93-9.

[88] Lee, KY; Peters, MC; Ennett, AB; Mooney, DJ. Controlled growth factor release from synthetic extracellular matrices. *Nature* 2000; *408*: 998-1000.

[89] Ajioka, I; Nishio, R; Ikekita, M; Akaike, T; Sasaki, M; Enami, J; Watanabe, Y. Establishment of heterotropic liver tissue mass with direct link to the host liver following implantation of hepatocytes transfected with vascular endothelial growth factor gene in mice. *Tissue Eng* 2001; *7*: 335-44.

[90] Yang, J; Chang, E; Cherry, AM; Bangs, CD; Oei, Y; Bodnar, A; Bronstein, A; Chiu, CP; Herron, GS. Human endothelial cell life extension by telomerase expression. *J Biol Chem* 1999; *274*: 26141-8.

[91] Yang, J; Nagavarapu, U; Relloma, K; Sjaastad, MD; Moss, WC; Passaniti, A; Herron, GS. Telomerized human micrivasculature is functional in vivo. *Nat Biotechnol* 2001; *19*: 219-24.

[92] Fontaine, M; Schloo, B; Jenkins, R; Uyama, S; Hansen, L; Vacanti, JP. Human hepatocyte isolation and transplantation into an athymic rat, using prevascularized cell polymer constructs. *J Pediatr Surg* 1995; *30*: 50-60.

[93] Homma, K; Himi, T; Hoki, K; Ezoe, K; Shintani; T; Yamaguchi, H; Fujita, T. A prefabricated oszeocutaneous flap for tracheal reconstruction. Plast *Reconstr Surg* 2003; *111*: 1688-92.

[94] Warnke, PH; Springer, IN, Wiltfang, J; Acil, Y; Eufinger, H; Wehmoller, M; Russo, PA; Bolte, H; Sherry, E; Behrens, E; Terheyden, H. Growth and transplantation of a custom vascularized bone graft in a man. *Lancet* 2004; *364*: 766-70.

[95] Edelmann, ER; Mathiowitz, E; Langer, R; Klagsbrun, M. Controlled and modulated release of basic fibroblast growth factor. *Biomaterials* 1991; *12*: 619-26.

[96] Laham, RJ; Sellke, FW; Edelman, ER; Pearlman, JD; Ware, JA; Brown, DL; Gold, JP; Simons, M. Local perivascular delivery of basic fibroblast growth factor in patients undergoing coronary bypass surgery: results of a phase randomised, double-blind, placebo-controlled trial. *Circulation* 1999; *100*:1865-71.

[97] .Hile, DD; Pishko, MV. Solvent-free protein encapsulation within biodegradable polymer foams. *Drug Deliv* 2004;*11(5)*: 287-93.

[98] Li, SH; Cai, SX; Liu, B; Ma, KW; Wang, ZP; Li, XK. In vitro characteristics of poly(lactic-co-glycolic acid) microspheres incorporating gelatin particles loading basic fibroblast growth factor. *Acta Pharmacol Sin* 2006; *27(6)*:754-9.

[99] Wissink, MJ; van Luyn, MJ; Beernink, R; Dijk, F; Poot, AA; Engbers, GH; Beugeling, T; van Aken, WG; Feijen, J. Endothelial cell seeding on crosslinked collagen: Effects of crosslinking on endothelial cell proliferation and functional parameters. *Thromb Haemost* 2000; *84*: 325-31.

[100] Pieper, JS; Hafmans, T; van Wachem, PB; van Luyn, MJ; Brouwer, LA; Veerkamp, JH; van Kuppevelt, TH. Loading of collagen-hepran sul fate matrices with bFGF promotes angiogenesis and tissue generation in rats. *J Biomed Mater Res* 2002; *62*: 185-194.

[101] Vailhe, B; Vittet, D; Feige, JJ. In vitro models of vasculogenesis and angiogenesis. *Lab Invest* 2001; *81*: 439-52.

[102] Nor, JE; Christensen, J; Mooney, DJ; Polverini, PJ. Vascular endothelial growth factor (VEGF)-mediated angiogenesis is associated with enhanced endothelial cell survival and induction of Bcl-2 expression. *Am J Pathol* 1999; *154*: 375-84.

[103] Nor, JE; Peters, MC; Christensen, JB; Sutorik MM, Linn S, Khan MK, Addison CL, Mooney DJ, Polverini, PJ. Engineering and characterization of functional human microvessels in immundeficient mice. *Lab Invest* 2001; *81*: 453-63.

[104] Joki, T; Machluf, M; Atala, A; Zhu, J; Seyfried, NT; Dunn, IF; Abe, T; Black, PM; Carroll, RS. Continous release of endostatin from microencapsulated engineered cells for tumor therapy. *Nat Biotechnol* 2001; *19*: 35-9.

[105] Mertsching, H; Walles, T; Hofmann, M; Schanz, J; Knapp, WH. Engineering of a vascularized scaffold for artificial tissue and organ generation. *Biomaterials* 2005; *26(33)*: 6610-6617.

[106] Ribatti, D; Nico, B; Vacca, A; Roncali, L; Dammacco, F. Endothelial cell heterogeneity and organ specificity. *J Hematother Stem Cell Res* 2002; *11(1)*: 81-90.

[107] Schultheiss, D; Gabouev, AI; Kaufmann, PM; Schlote, N; Mertsching, H; Haverich, A; Stief, CG; Jonas, U. [Biological vascularized matrix (BioVaM): a new method for solving the perfusion problems in tissue engineering]. *Urologe* 2004; *43(10)*: 1223-8.

[108] Hodde, J. Naturally occurring scaffolds for soft tissue repair and regeneration. *Tissue Engineering* 2002; *8*: 295-308.

[109] Badylak, S; Liang, A; Record, R; Tullis, R; Hodde, J. Endothelial cell adherence to small intestinal submucosa: an acellular bioscaffold. *Biomaterials* 1999; *20*: 2257-2263.

[110] Voytik-Harbin, SL; Brightman, AO; Kraine, MR; Waisner, B; Badylak, SF. Identifikation of extractable growth factors from small intestinal submucosa. J. Cell. *Biochem.* 1997; *67*: 478 – 491.

[111] Lantz, GC; Badylak, SF; Hiles, MC; Coffey, AC; Geddes, LA; Kokini, K; Sandusky, GE; Morff, RJ. Small intestinal submucosa as vascular graft. *A review. J. Invest. Surg.* 1993; *6*: 297-310.

[112] Badylak, S; Obermiller, J; Geddes, L; & Matheny, R. Extracellular matrix for myocardial repair. *Heart Surg Forum* 2003; *6*: E20-26.

[113] Tucker, O.P; Syburra, T; Augstburger, M.; van Melle, G; Gebhard, S; Bosman, F; & von Segesser, L.K. Small intestine without mucosa as a growing vascular conduit: a porcine experimental study. *J Thorac Cardiovasc* Surg 2002; *124*: 1165-1175.

[114] Record, R.D; Hillegonds, D; Simmons, C; Tullius, R; Rickey, F.A; Elmore, D; & Badylak, S.F; In vivo degradation of 14C-labeled small intestinal submucosa (SIS) when used for urinary bladder repair. *Biomaterials* 2001; *22*: 2653-2659.

[115] Atala, A. Future perspectives in bladder reconstruction. *Adv Exp Med Biol* 2003; *539*: 921-940.

[116] Musahl, V; Abramowitch, S.D; Gilbert, T.W; Tsuda, E; Wang, J.H; Badylak, S.F, & Woo, S.L. The use of porcine small intestinal submucosa to enhance the healing of the medial collateral ligament--a functional tissue engineering study in rabbits. *J Orthop Res* 2004; *22*: 214-220.

[117] Huber, J.E; Spievack, A; Simmons-Byrd, A; Ringel, RL; & Badylak, S. Extracellular matrix as a scaffold for laryngeal reconstruction. *Ann Otol Rhinol Laryngol* 2003; *112*: 428-433.

[118] Macchiarini, P; Walles, T; Biancosino, C; & Mertsching H. First human transplantation of a bioengineered airway tissue. *J Thor Cardiovasc Surg* 2004; *128*: 638-41.

[119] Badylak, S.F. The extracellular matrix as a scaffold for tissue reconstruction. *Semin Cell Dev Biol* 2002; *13*: 377-383.

[120] Biancosino, C; Zardo, P; Walles, T; Wildfang, I; Macchiarini, P; Mertsching, H. Generation of a bioartificial fibromuscular tissue with autoregenerative capacities for surgical reconstruction. *Cytother* 2006; *8*: 178-83.

[121] Johnson, JI; Decker, S; Zaharevitz, D; Rubinstein, LV; Venditi, JM; Schepartz, S. Relationship between drug activity in NCI preclinical in vitro and in vivo models and early clinical trials. *Br J Cancer* 2001; *84*: 1424-1431.

[122] Walles, T; Weimer, M; Linke, K; Michaelis, J; Mertsching, H. The potential of bioartificial tissues in oncology research and treatment. *Onkologie* 2007; *30*: 388-94.

INDEX

A

B

C

D

E

F

G

H

N

O

S

Q

R

T

U

V

W

X

Y